MW01611458

THE
SAVIOR

THE
SAVIOR

GENERAL OLIVER PRINCE SMITH

Nick Ragland
Tom Schwettman

ORANGE *frazer* PRESS
Wilmington, Ohio

ISBN 978-1939710-444
Copyright©2016 Nick Ragland and Tom Schwettman
All Rights Reserved

Published for the author by:
Orange Frazer Press
P.O. Box 214
Wilmington, OH 45177

Telephone: 937.382.3196 for price and shipping information.
Website: www.orangefrazer.com

Book and cover design: Alyson Rua and Orange Frazer Press

Library of Congress Control Number: 2016906161

For Marty and Patti

"When the Korean War broke, somewhat less than ten percent of the small United States Marine Corps had seen combat. But fortunately for the Corps, the percentage was highly concentrated within officer and key NCO grades; most of the Marine troop leaders knew what war was like...

...And these men walked with a certain confidence and swagger. They were only young men like those about them in Korea, but they were conscious of a standard to live up to, because they had good training, and it had been impressed upon them that they were United States Marines...

...These Marines had pride in their service, which had been carefully instilled in them, and they had pride in themselves, because each man had made the grade in a hard occupation. They would not lightly let their comrades down...

...Marine human material was not one whit better than that of the human society from which it came. But it had been hammered into form in a different forge, hardened with a different fire. The Marines were the closest thing to legions the nation had."

–Colonel T.R. Fehrenbach, USA (Ret.)
Excerpts from This Kind of War

ACKNOWLEDGMENTS

We would like to thank the following individuals for their guidance and assistance in constructing this novel: Patty Hogan, our editor, for her thoughts, valuable comments and suggestions. We can't overstate our debt to Patty. Her husband and son served in the Marine Corps.

Lieutenant Colonel Charles Neimeyer (Retired), Director of the Historical Division and the Gray Research Center at Marine Corps University for his help and guidance and for granting us access to General Smith's records.

To Joe Zeinner who designed the cover and all the maps.

To Marcy and Sarah Hawley at Orange Frazer Press.

And last but not least, Mary Miller, who was the first to read, type, correct, and make comments on the manuscript.

CONTENTS

LIST OF MAPS

ORGANIZATION OF THE 1ST MARINE DIVISION* (approximate figures)	
Squad	12 Marines
Platoon	50 Marines
Company	200 Marines
Battalion	800 Marines
Regiment	3,000 Marines
Division	20,000 Marines

INFANTRY REGIMENTS OF THE 1ST MARINE DIVISION	
1st Marine Regiment	(1st Marines)
5th Marine Regiment	(5th Marines)
7th Marine Regiment	(7th Marines)

BATTALIONS OF THE 1ST MARINE REGIMENT	
1st Battalion, 1st Marines	(1/1)
2nd Battalion, 1st Marines	(2/1)
3rd Battalion, 1st Marines	(3/1)

BATTALIONS OF THE 5TH MARINE REGIMENT	
1st Battalion, 5th Marines	(1/5)
2nd Battalion, 5th Marines	(2/5)
3rd Battalion, 5th Marines	(3/5)

BATTALIONS OF THE 7TH MARINE REGIMENT	
1st Battalion, 7th Marines	(1/7)
2nd Battalion, 7th Marines	(2/7)
3rd Battalion, 7th Marines	(3/7)

*General Oliver Prince Smith served with the 1st Marine Division both in World War II and Korea.

PREFACE

This is a work of historic fiction about Marine Major General Oliver Prince Smith who commanded the 1st Marine Division during the epic battle of Chosin Reservoir. This battle was fought in late November and early December of 1950, in sub-zero weather in the mountains of North Korea. This story is told through the voices and courageous actions of the men of this legendary division. Fourteen Medals of Honor were awarded to Marines for actions that pertain to the Chosin Reservoir campaign. Seven of these medals were awarded posthumously.

What most Marines may have forgotten, or maybe never knew, is the name of the man who commanded this division and who saved it from almost certain annihilation.

1

THE GENERAL

The car looked as pedestrian as its owner. It was a two-toned green sedan with plenty of miles and years on it. Nothing flashy here. But like the owner, the car was always reliable.

It wasn't until the car pulled up to the gate at the Marine Corps base at Camp Pendleton, California, that the two stars became visible. The sentry immediately snapped to attention and saluted and managed to say, "Good morning, sir."

The white haired man in the car replied, "Good morning, corporal. It looks like another fine summer day."

"Yes sir," the sentry responded. "Can I help you with any directions?"

"No, corporal. I think I know where I am going," responded the driver.

It was a hot, sunny morning on July 26, 1950. The heat didn't bother the car's occupant as he had been raised in California.

The general drove through the gate and onto the base. As he went down the road, he passed hundreds of Marines marching along the side of the road. The southern California sunshine lit the brown mountains and hills that ringed the road. The young faces and the cadence of the march took him back over thirty years earlier to 1917 when he became a Marine. He had joined to fight the Huns but World War I had passed him by. He spent the war years in garrison duty on Guam. This was followed by barracks duty, sea duty and staff duty; then three years in

Haiti, starting in 1928. He had attended military schools and he had taught in them. Some Marines referred to him as "the Professor." His friends called him "O.P." It wasn't until World War II that he finally saw the action that Marines typically craved. He had fought that war in the Pacific. Now there was another war.

He pulled up to a drab, wooden building and parked in front of it. There was a sign that read: Commanding General, 1st Marine Division. He stopped for a moment to look at the sign and then he climbed out of the car, stood ramrod straight and walked into the building. As he entered the lobby, he turned and walked into an office and said to the young Marine at the front desk, "Good morning corporal, I'm General Smith."

All the Marines in the office immediately stood and one of the men said, "Welcome aboard, sir. We've been expecting you. Can I show you to your office?"

"I'd appreciate that," replied Smith.

Major General Oliver Prince Smith was fifty-six years old. After thirty-three years in the Marine Corps, his moment had finally arrived.

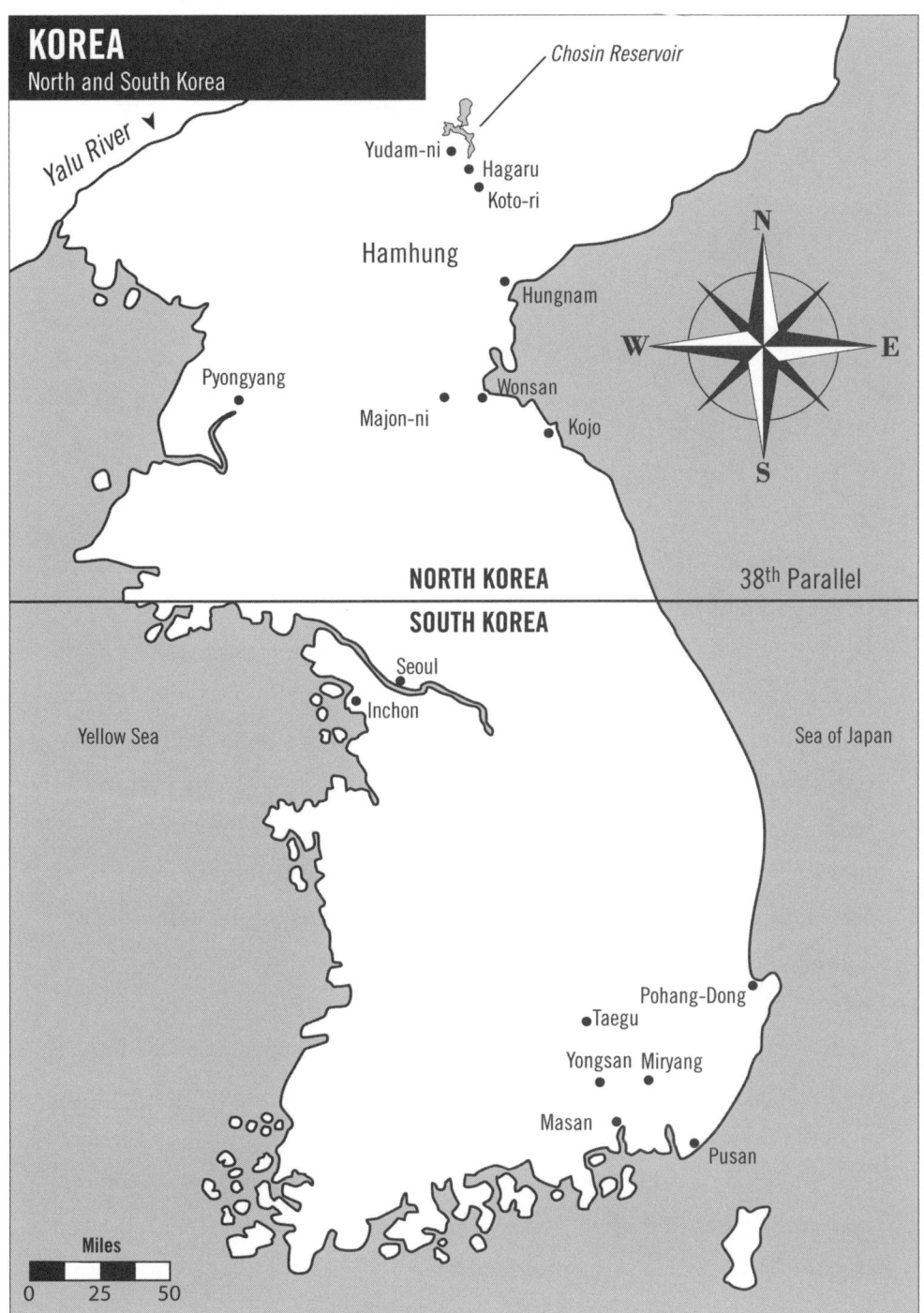

KOREA
North and South Korea

Chosin Reservoir
Yalu River
Yudam-ni
Hagaru
Koto-ri
Hamhung
Hungnam
N
Pyongyang
Wonsan
W   E
Majon-ni
Kojo
S

NORTH KOREA   38th Parallel
SOUTH KOREA

Seoul
Inchon
Yellow Sea   Sea of Japan

Pohang-Dong
Taegu
Yongsan Miryang
Masan
Pusan

Miles
0   25   50

2

ROOTS OF THE KOREAN WAR

Prior to World War II, the Korean peninsula had been under Japanese control for thirty-five years. During this time Koreans were forced to serve Japan in many menial and degrading capacities. When the atom bomb brought Japan to absolute defeat, ending their role in World War II, there was a rush to claim Korea and all the other territories held by Japan. The Russians belatedly declared war on Japan and invaded the Korean nation taking all Korean territory north of the 38th parallel. The newly formed United Nations, led by the United States, took everything south of this marker line thereby separating the Korean Peninsula, which for centuries had been a unified country, into two distinct nations: North Korea, following communism and South Korea becoming a democracy. The U.S. didn't put up much of a fight for Korea. Their focus was on Europe and the post war divisions occurring there. They were satisfied with getting Seoul, the traditional capital of the old unified Korea, as well as Inchon and the Pusan ports. These areas had greater economic benefits than most of the land above the 38th parallel.

This post World War II division arbitrarily achieved by agreement between Russia and the U.S. was never a peaceful one. Kim Il-Sung became North Korea's first premier in 1948 and soon made his intentions known that he planned to reunite the two Koreas into one communist nation. He didn't wait long to act on his plan.

The Korean War started on Sunday, June 25, 1950. That's the day when most wars start, when the army with the rest of the world is fast asleep. A quiet Sunday morning was when the North Korean army with 90,000 soldiers swarmed across the 38th parallel from communist North Korea into democratic South Korea. Funded and supported by both China and Russia, the North Koreans vastly outmatched their southern brothers. They easily defeated the estimated 38,000 South Korean troops on the frontier, forcing them to retreat south. The South Korean troops were badly trained and poorly armed. Some had been gang pressed into military service. Overwhelmed by North Korean tanks and troops, Seoul, the capital city of South Korea, fell in just three days.

On June 27 the United Nations Security Council voted to authorize military action for the first time in its short history. Russia could have vetoed the vote in the Security Council since the U.S., Russia, Nationalist China, Great Britain, and France each had absolute veto power. But, Russia had boycotted the Security Council meetings on Korea because they were protesting the fact that the Chinese Nationalists were still seated on the Council rather than the Chinese Communists. With Russia not present for the vote, the Security Council authorized sending a predominately American force, under the UN flag, to fight in Korea.

U.S. forces, under the command of General Douglas MacArthur were ordered into battle by President Harry Truman. MacArthur, who was in charge of U.S. forces in the Far East, had been a national hero in World War II, but many who dealt with him found him to be overly confident and egotistical. His thirty-two years as a general had done nothing to temper his vanity. At age seventy, MacArthur was now acting as Supreme Commander for Allied Powers from his office in Tokyo.

The declaration of war forced U.S. Army soldiers to leave their comfortable jobs as occupying troops in Japan. A new post-war Army soldier, soft, untrained and undisciplined, began arriving in Korea on July 1. These troops were not ready for combat and were no match for the

North Korean army. On that same day, MacArthur requested a Marine regiment with a supporting air group. Brigadier General Edward Craig, the Assistant Division Commander of the 1st Marine Division would command the 1st Marine Brigade, made up of the 5th Marine Regiment and Marine Aircraft Group 33. They would sail for Korea on July 12.

Craig was a seasoned combat veteran. He was an Army brat, and planned to make the Army his career. But in order to get a commission in the Army to serve in World War I, he needed to be twenty-one years old. The Marines gave the commission at age twenty, so the twenty-year-old Craig joined the Marines. His Army father thought this was a terrible mistake but Craig proved him wrong.

In World War II, Craig was awarded the Bronze Star while commanding the 9th Marines at Bougainville and then in the recapture of Guam he received the Navy Cross. While commanding the 1st Provisional Marine Brigade in the Pusan Perimeter at the start of the Korean War, he was awarded the Distinguished Service Medal and the Silver Star.

MacArthur next asked for a Marine division. In response, President Truman authorized the activation of the Marine Corps Reserve on July 19.

3

THE 1ST MARINE DIVISION

In late July, Major General Oliver Prince Smith took command of the 1st Marine Division. Smith was not new to the division. He had served with it in World War II in some of the bloodiest battles of the war. He commanded the 5th Marines in the Talasea phase of the Cape Gloucester battle. He was also the assistant division commander for the bloody battle of Peleliu and he finished the war as the Marine Deputy Chief of Staff of the 10th Army on Okinawa. On Cape Gloucester, Smith was awarded the Bronze Star Medal, on Peleliu the Legion of Merit, and on Okinawa a second Legion of Merit. He was experienced in combat and had been well decorated.

Smith was far from the stereotype of what most people thought a Marine Corps general should be. Smith was a gentleman, both intelligent and reserved, with a high degree of integrity. Raised as a Christian Scientist, he did not drink and never used profanity. He shunned publicity. He didn't yell or scream and he rarely raised his voice. He remained faithful and devoted to his wife despite long and difficult separations. He was a compulsive note taker and always had a notebook with him. He was a perpetual student, and well-read in military history. However, he loved the rough and tough comradery of the Marine Corps. Beneath his calm demeanor, lay a strong and competent leader.

In 1948, General Smith became the Assistant Commandant of the Marine Corps. After his two year tour in Washington D.C. in that post, he was rewarded with the command of the 1ˢᵗ Marine Division.

The 1ˢᵗ Marine Division was the Corps' oldest and most famous division. During World War II, it had been used in the Pacific theater with legendary results. It was the 1ˢᵗ Marine Division that had fought and won the first offensive land battle of World War II at Guadalcanal. It went on to fight on Cape Gloucester followed by the battle of Peleliu and finished the war in the home islands of Japan with the capture of Okinawa. General Smith had played his part in the Marine Corps' battles in the Pacific.

A Marine division is traditionally formed on the concept of threes: three infantry regiments form the backbone of a division. Each regiment consists of three infantry battalions. Each battalion has three infantry companies. And each company has three infantry platoons and finally each platoon has three squads of thirteen men each. In addition, the division is supported by units of artillery, tanks, engineers, medical and service and support groups.

When General Smith took command of the 1ˢᵗ Marine Division, it was a shell of its former self. Within months after the close of World War II, the Marine Corps' strength was drawn down drastically, from six divisions to two divisions and from a peak of 485,000 Marines to 156,000. By February 1948, it was further reduced in size to just 92,000. Infantry companies had been slashed from three platoons to two and infantry battalions from three companies to two. Some wondered if the Marine Corps was destined to fade into history. When General Smith assumed command of the 1ˢᵗ Marine Division, he had only 3,500 men. Smith knew of the depleted forces from his tour in Washington, but seeing the numbers in a briefing book and having to go to war with so few men were two very different things. Smith's job was to quickly bring the division up to the fighting strength of over 22,000 men.

Smith worked furiously for the few weeks he had at Camp Pendleton. Daily reports from Korea brought disheartening news of the state of the U.S. Army and the South Korean forces. Smith knew the troops he was assembling would not have long to become battle ready. Forming a Marine division was a daunting task. Men had to be found, equipped, trained and transported. There weren't nearly enough Marines at Camp Pendleton to accomplish this. Many that had been there were already en route to Korea with General Craig's 1st Marine Brigade.

General Smith's ranks were immediately swelled when President Truman called up the Marine reserves in July. But even all the reservists were not nearly enough to man the 1st Division. Men on sea duty, recruiting duty and in other units at Camp Lejeune on the east coast were brought in to fill the ranks of the division.

Smith also had to work on the logistics of arming and supplying his men. He needed everything from toilet paper to tanks. Old, refurbished tanks and rifles from World War II were assembled. Unfortunately many of these had been in moth balls too long and didn't work. Smith and his people had to scramble to find equipment and arms that did.

While assembling Marines and equipment was vital, equally important would be the task of putting together his staff. Smith knew that the young men heading to Korea lacked combat experience and, in many cases, even the basic Marine training that was critical to performance in battle. Some of these reservists had never been to boot camp. They had escaped the dreaded Parris Island training with only weekend warrior experience of a reservist plus a couple of weeks a year of Marine training. To compensate for the lack of veterans in his ranks, Smith knew he had to find combat experienced officers to fill his staff.

Smith's division staff was in many respects like family. Some of the members of his staff, like his assistant division commander Brigadier General Eddie Craig, had been inherited from the previous commander. Smith had first met Craig in 1937 when Craig was a student in the senior course at Marine Corps School in Quantico and Smith was the Infantry Tactics Instructor.

The duties of the assistant division commander are to be the commander's right hand, anticipating needs and carrying out his orders. Smith would give Craig a lot of responsibility and the authority to make decisions when Smith wasn't there. Craig was already in Korea, commanding the 1st Marine Brigade, when Smith took command of the 1st Marine Division. Smith knew that he and Craig shared the same values. He also knew Craig was a commander who put himself in the thick of action. He ate with his infantrymen and slept on the ground with them. Smith was told that before leaving California, Craig had told his 1st Marine Brigade that he would never order them to retreat but would ask they fight as Marines have always fought. He could rely on Craig for a good assessment of the military and political situation he was about to enter.

Unlike the inherited Craig, Smith's G-3 (operations officer) Colonel Alpha Bowser had been personally requested by the general. Bowser was originally from Pennsylvania and had graduated from the Naval Academy. He opted for a commission as a 2nd lieutenant in the Marines rather than a career in the Navy. Bowser had a strong reputation as an instructor, staff officer, and expert in naval gunfire. He commanded an artillery battalion on Iwo Jima during World War II and always thought of himself as an artillery man. He was awarded two Bronze Stars and a Legion of Merit. Marines who knew Bowser well used one word to describe him: "brilliant." Bowser would be responsible for the planning and operations of the division. If Craig was Smith's right hand, Bowser would be his left.

The fighting force of Smith's 1st Marine Division was made up of his three infantry regiments. Commanding the 1st Marine Regiment was Colonel Lewis B. "Chesty" Puller. In the early 1920s, this legendary Marine had spent five years fighting the rebels in Haiti and then in the late 20s and early 30s, he fought the bandits in Nicaragua where he earned two Navy Crosses. During World War II, he was awarded his third Navy Cross during the battle for Guadalcanal and at Cape Gloucester a fourth Navy Cross. He finished the war as the Commanding Officer of the 1st Marines in the costly battle for Peleliu. Puller's style was to lead from the front and he had more combat experience than the other two regimental commanders.

Smith and Puller had known each other since 1931 when they both attended military schools at Fort Benning, Georgia: Smith attending the Field Officer School and Puller, the Company Officer School. They had also served together on Peleliu during World War II. These two men could not have been more different in their appearance or their personalities. Smith was tall and slender with a quiet and distinguished bearing while Puller was short, profane and one of the most colorful officers in the Marine Corps. Because of his barrel chest, most Marines referred to Puller as "Chesty" but no one ever called him that to his face. He was always called Lewis or Lewie by his peers. Smith called him Lewie and knew that in Puller he had a leader that men would follow anywhere. Chesty Puller was known to all as a "Marine's Marine."

Lieutenant Colonel Raymond L. Murray would command the 5th Marine Regiment. Murray was a graduate of Texas A&M. As a major on Guadalcanal during World War II, he was awarded his first Silver Star. On Tarawa came a second Silver Star. On Saipan, where he was wounded, he was awarded a Navy Cross and a Purple Heart. While Smith was working in California, Murray was already in Korea, serving in the 1st Marine Brigade under General Craig. While in command of the 5th Marines on the Pusan Perimeter at the start of the Korean War, he was awarded his third and fourth Silver Stars. It was unusual to have a lieutenant colonel in command of a regiment, but his outstanding combat record kept him in a leadership position. Murray was a young, hard-charging leader and was another person Smith would rely on for his combat experience.

Colonel Homer L. Litzenberg Jr. led the 7th Marine Regiment. During World War II, he commanded the 3rd Battalion 24th Marines and became the Regimental Executive Officer for the assault of Roi-Namur in the Marshall Islands. When the Korean War started, he was in command of the 6th Marines at Camp Lejeune. This regiment was sent to Camp Pendleton and became the 7th Marines. Litzenberg had the least combat experience of the regimental commanders but this would soon be remedied. He was a forceful commander with a reputation of being tough on subordinates. General Smith had heard that Litzenberg might be difficult to get along with, but his impressive resume won him the job. Smith also had the opportunity to work with "Litz" at Pendleton where Litzenberg was tasked with reactivating the 7th Marines and having them ready to sail for Korea by the first of September. He had seen the man in action and knew he could depend on him.

Smith's three infantry regiments were very well led.

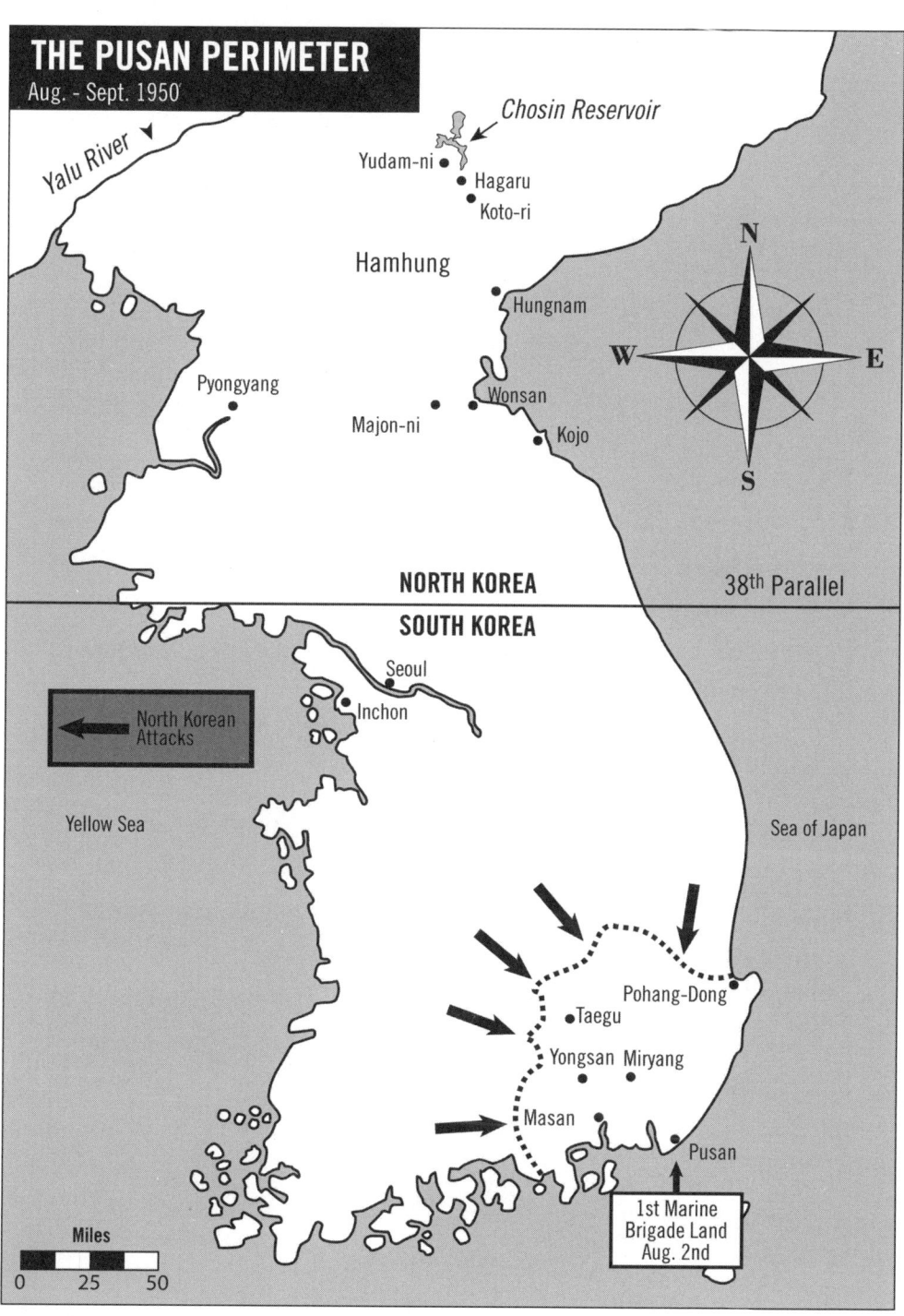

THE PUSAN PERIMETER
Aug. - Sept. 1950

Yalu River

Chosin Reservoir

Yudam-ni

Hagaru

Koto-ri

Hamhung

Hungnam

Pyongyang

Majon-ni

Wonsan

Kojo

N
W **E**
S

NORTH KOREA

38th Parallel

SOUTH KOREA

Seoul

Inchon

North Korean
Attacks

Yellow Sea

Sea of Japan

Pohang-Dong

Taegu

Yongsan Miryang

Masan

Pusan

1st Marine
Brigade Land
Aug. 2nd

Miles

0   25   50

4

PUSAN, INCHON, AND SEOUL

By August 1, U.S. Army forces had been pushed back to an area in southeast Korea known as the Pusan Perimeter. This was a disastrous start for an army that just five years earlier had achieved a global triumph in World War II. The American public, as well as President Truman, largely wrote off Korea as a "police action" which lacked popular support. The U.S. Army had also significantly deteriorated in strength, numbers and leadership since the end of the great war. The military's focus since 1948 had been on NATO and the growing threat of communism in Russia and Eastern Europe. Asia had been largely neglected. But when faced with the grim news that American forces were losing to a North Korean army, military resources had to be diverted to Asia. America's reputation could not stand that kind of setback.

The same was true of intelligence forces. The newly formed CIA focused on Europe and had few resources in Asia. They had failed miserably with respect to anticipating the Korean War, not knowing of the North Korean invasion plans and having little to no intelligence on the Chinese strategy related to Korea.

On August 2, after three weeks at sea, the 1ˢᵗ Marine Brigade, led by Brigadier General Eddie Craig, landed at Pusan, a major port city on the southeast coast of South Korea. Five days later they went into action. These Marines from Camp Pendleton would soon be called the

"Fire Brigade" and would help stop the North Korean Army and save the Pusan Perimeter. Marines in the Fire Brigade had to be everywhere at once to back up U.S. Army and South Korean troops that were taking a beating.

General MacArthur had conceived a bold plan to reverse the tide of this war with an amphibious landing behind enemy lines at the port of Inchon. This port was located twenty miles west of Seoul and at the opposite end of South Korea from Pusan. If MacArthur's plan succeeded, the landing would isolate the North Korean army, cut their supply lines and then put the Marines in position to trap the invaders from the north. It would also help the UN forces to "liberate" Seoul, providing a great psychological as well as a military victory.

Early on the morning of August 22, Smith arrived in Japan and was met by Admiral Doyle. The two old friends shook hands and Admiral Doyle said, "O.P., welcome to Tokyo."

The Marine general replied, "Thanks Jimmy. Good to see you. It was a long trip from the States."

They climbed into the Navy gray staff car and headed toward the pier. On the way, the admiral shared with Smith that the Marines were going to land at Inchon on September 15. It was the first that Smith had heard of this and he was stunned.

"I know that you're meeting with MacArthur this afternoon and I wanted to give you a heads up. I have real concerns about this landing. I think it's risky as hell. Unfortunately, O.P., it is not my call. It's MacArthur's," said Admiral Doyle. After a short drive, they arrived at the ship.

Smith's new quarters were aboard the USS Mount McKinley, Doyle's command ship, which was anchored in Tokyo Bay. This ship had served as a flag ship in the battles of Peleliu, Leyte and Okinawa. The USS Mount McKinley would also serve as MacArthur's headquarters during the upcoming Inchon landing.

Smith had closed his command post at Camp Pendleton on the eighteenth and had been flown to Japan. As is often the case, the summer weather in Japan was hot and very humid. The ship's quarters were not air conditioned and Smith found it hard to stay awake in the close quarters of the cabin after the long flight and the time change. This cabin would be his home for the next three weeks.

After lunch, Colonel Bowser met with Smith in his quarters. They exchanged pleasantries and did a little catching up before the operations officer got to the point. "Sir, we are going to land at Inchon on September 15."

Smith replied, "Admiral Doyle told me about this plan when he brought me over from the air field. He was very skeptical about it. What can you tell me?"

"Sir, the plan seems set." Bowser pulled out a map and showed it to Smith. "It is General MacArthur's baby. Inchon is on the west coast of Korea close to Seoul and not far from the 38th parallel."

"No wonder MacArthur was insistent on getting Marine support, especially if he wants to do an amphibious landing," said Smith.

Bowser and Smith were both well versed in amphibious landings as was General Craig. After World War II, Smith had taught and written a great deal about the strategy of such landings. He was known as something of an expert in this area. Craig had planned and participated in the amphibious landing at Iwo Jima and had also been a chief instructor training troops and amphibious forces.

"Will we have time to train or rehearse for the landing?" asked Smith.

"No sir! MacArthur wants this invasion to take place on September 15. Puller's 1st Marines are due to arrive in Japan around the 1st of September and then still have to get to Inchon. Murray's 5th Marines are still fighting in the Pusan Perimeter. Hell, sir, MacArthur has to release them from that battle and we still have to get them up to Inchon on the other side of the country. Litzenberg's 7th Marines won't arrive from California in time for the landing."

"Who will command this operation?"

"Well sir, I assume it's going to be General Shepherd. I was told that the 1st Marine Division and the Army's 7th Infantry Division will form a new unit called X Corps. MacArthur decided that the nature and location of the planned Inchon landing meant that it should be directed by a command separate from the 8th Army. He then created X Corps."

Both men knew Marine Lieutenant General Lem Shepherd. He was an outstanding, highly decorated Marine from World War I and World War II and now commanded the Fleet Marine Force Pacific and was in line to be the next Commandant of the Marine Corps. Shepherd had commanded everything from a platoon to a division and technically he was Smith's immediate superior. Both Smith and Bowser took comfort in the fact that Shepherd would probably lead this action.

For the next two hours, Bowser briefed Smith on everything he knew about the port of Inchon. Smith filled a small notebook during the briefing. The information did nothing to allay Smith's concerns about the landing, but it better prepared him for his upcoming meeting with the Supreme Commander. The two men also discussed some potential alternate sites for a landing.

Smith and Bowser talked for a while longer and then the general looked at his watch and said, "I better get going. I have a meeting with General MacArthur at 5:30 and I don't want to be late. I have no idea what Tokyo traffic is like. You know it's strange to be here in Japan without being at war with them. Just five years ago they were the enemy."

Smith packed up his notes and maps and was taken ashore by a tender to where a car was waiting for him.

MacArthur's headquarters were in the Dai Ichi building. Pre-war, this stately structure was the headquarters of an insurance company and was one of the few high rise buildings not bombed out in Tokyo. Its location across from the Imperial Palace saved it. Smith's first impression when he entered the building was that it seemed more like a palace than a headquarters. The place crawled with spit and polish soldiers who were armed with bayoneted rifles. The afternoon traffic had not been as bad as expected. When Smith got off the elevator on the sixth floor, he was happy to note that he was early for the scheduled 5:30 meeting. His mood altered however as he cooled his heels in the lobby until 7:00 pm. When his name was finally called nearly two hours after his arrival, he was ushered into the office of Army Major General Ned Almond, not General MacArthur.

"General Smith, I'm General Almond, General MacArthur's Chief of Staff. Come on in my office." Almond motioned for Smith to take a seat and then he sat down behind the desk.

Almond started in, "Son, I assume you know that the plan is to have the 1st Marine Division land at Inchon on September 15."

Smith had contained his anger about the delay for which Almond offered no apology, but had more trouble swallowing this new insult. "Son" was a term that an older officer might use in talking to a much younger enlisted man. But Smith was fifty-seven years old and the same rank although, he was only one year younger than Almond. He was furious but he wasn't about to take part in Almond's game of one upmanship.

Smith controlled his exasperation and started to explain all the difficulties of this planned amphibious operation. "I have several concerns about this landing. While I just learned about it, what I do know is that it poses real risks. First of all, there are no beaches at Inchon, only seawalls and mud flats. Second, the channels are narrow and it is typhoon season. Third, the tides can be thirty-two feet or more and so the landing would have to take place late in the day. We will need ladders to scale the sea walls making our men sitting ducks for enemy fire. Once ashore, this would give us little time to get organized before it gets dark. But more important, we won't have any time to practice for such a risky landing. As you know, most of the 1st Marine Division isn't in Korea yet.

"And, General Almond, the area has a population center of about 250,000 Koreans. I have been working with my operations officer and we have identified a couple of alternative sites that would not pose all of these problems," said Smith.

Almond seemed annoyed at these comments and replied, "Son, these difficulties you refer to I'm sure are purely mechanical. General MacArthur has selected Inchon and Inchon it will be."

The second "son" set Smith's teeth on edge. He assumed that Almond's patronizing attitude was a deliberate attempt to antagonize him. It soon became apparent to Smith that Almond, in addition to being rude and condescending was also not a good listener. And perhaps, most alarmingly, not a smart tactician.

While Smith made another effort to convey his concerns, Almond abruptly stood up and said, "It's time to meet the general."

Smith followed Almond into General MacArthur's office.

The Supreme Commander was standing looking out the window when Smith came in. He beckoned him over and warmly shook his hand.

"See, General Smith, the garden over there?" asked MacArthur.

"Yes sir," responded Smith.

"That's the Imperial Garden where good old Tokyo Rose used to tell everyone that I would be publicly hanged," said MacArthur.

MacArthur then told both men to have a seat. The general sat behind his desk and started to pour on the charm. He didn't ask for Smith's assessment of the Inchon landing. Others had obviously already conveyed Smith's same concerns. Smith knew from his friend Admiral Doyle that the Navy shared his opinions on the Inchon site. MacArthur knew there were lots of objections to the Inchon operation but he was not a man to have his decisions challenged. He had made up his mind and had all the arguments on why the landing would succeed.

"General Smith, Inchon is the last place the North Koreans would expect us to land because of the obvious difficulties with the site. The

1ˢᵗ Marine Division is experienced in these types of landings and I personally asked for your division. This landing will be decisive. We'll take the North Koreans by surprise, then we'll move on and capture Seoul. As you know, Seoul is just spitting distance from Inchon."

MacArthur stood up and pointed to a large map of Korea on the wall behind him.

"When we do this, we'll cut off the North Korean supply lines from the north and we'll trap their army in the south between your Marines and the 8ᵗʰ Army."

MacArthur sat down and leaned back in his chair. He took his corncob pipe, stuffed it full of tobacco and lit it. Smith thought the general looked his seventy years. His office was not as regal as Smith assumed it would be in this palatial building. It was not a corner office and wasn't ornate although it did overlook the Imperial Gardens. Smith noticed the general's desk had no drawers. He also saw a pair of reading glasses on MacArthur's desk. He remembered a combat photographer telling him that no photograph of the general could be taken when he was wearing his glasses. If this wasn't such a serious meeting, Smith would have smiled.

After taking a couple of pulls on his pipe, the general continued. "As you know, there are people in Washington who want to get rid of the Marine Corps. It's no secret that President Truman never liked the Marine Corps and that Generals Eisenhower and Bradley would like to downsize the Corps to regimental size. Eisenhower doesn't want two "land armies" and as I recall General Bradley said something to the effect that the atomic bomb would make amphibious landings obsolete. The Corps has been downsized so often that it is getting close to extinction."

Then MacArthur leaned forward and gave Smith a very compelling reason to be positive about the Inchon landing. "General Smith, you land at Inchon and you take Seoul, I can assure you that you will have secured the future of the Marine Corps. I know this operation will be sort of helter skelter but this landing will win the war and the troops will be home by Christmas. You will show America that we still need our Marines and that amphibious landings still have a place in battle plans. Your 1st Marine Division is going to win the war by landing at Inchon."

MacArthur stood up. The meeting was over. It was a one-half hour monologue. Smith felt MacArthur was very dramatic and if he hadn't been a soldier, he would have made a great actor. He also thought that his unmistakable sonorous voice which he had heard many times on the radio was as strong as ever. But Smith knew it would take a lot more than a radio voice to make the landing at Inchon a success.

Smith left MacArthur's headquarters and returned to the command ship. He knew that his concerns about the landing at Inchon had been disregarded. MacArthur's decision was final!

He wasn't given the opportunity to voice his objections to MacArthur and had little faith that Ned Almond would be sharing these objections with the Supreme Commander. Smith gave this news to Bowser when he got back to the ship. "Al, we need to start planning immediately for this invasion," said Smith.

"I'm already on it," replied Bowser.

Smith's team began working in earnest for the Inchon landing. Maps, charts and notes quickly filled the small cabin assigned to Smith.

The Inchon landing's code name was operation Chromite. It was opposed by the Joint Chiefs in Washington who believed that it left the United States without any reserve troops if anything went wrong. MacArthur convinced them to let it go forward, but he knew that if it failed, the failure would be clearly attributed to him.

MacArthur's invasion planning meeting of August 23 included his staff, but neither General Smith nor General Shepherd were invited. General Almond later met with Smith and Army General David Barr. At the meeting, Smith again proposed alternate landing sites for the invasion and again these suggestions were summarily dismissed.

On the next day, Smith met with Shepherd who promised to meet with MacArthur to try to convince him that alternative sites had real advantages. Smith fervently hoped that hearing from a real expert in amphibious landings would persuade MacArthur, but it did not.

Colonel Bowser appeared at Smith's cabin door four days later on August 26 with a grim expression.

"General, I just learned that Lem Shepard won't be commanding X Corps," said Bowser.

"You're kidding," replied Smith. "Lem knows more about amphibious landings than just about anyone on the planet. So who got the command?"

Bowser put his head down and the expression on his face told Smith that he wasn't going to like the answer.

"It's General Almond," said Bowser.

"Is he quitting as MacArthur's Chief of Staff to take this on?" asked Smith.

"No," Bowser replied. "Word is he is staying on as Chief of Staff."

"How can he do both? And what does he know about amphibious landings?" asked Smith.

"Permission to speak frankly, sir?"

"Sure, I need your unvarnished opinion here Al," replied Smith.

"The word I am hearing, General, is that MacArthur is known for giving his cronies commands not based on their competency or skills, but because he thinks they will always be loyal to him personally," said Bowser.

"I've heard that same thing before," said Smith, "but I really never believed he would want a sycophant going into battle in place of a genuine leader. I remember reading or hearing once that General Marshall met with MacArthur during World War II and told him, 'General, you don't have a staff, you have a court.' I'm afraid Marshall might have hit the mark."

"You know Eddie Craig is going to blow a gasket when he hears about this," Smith told Bowser. "Craig replaced Shepherd as commander of the 9th Marines in 1943 and I've often heard him call Shepherd an amphibious assault genius."

Almond, the Army general with no amphibious experience, was a Virginian and had served in the final months of World War I. In World War II, he commanded the Army's 92nd Infantry Division, an all-African-American unit in a segregated Army. The division performed poorly in combat and Almond blamed its poor performance directly on the African-American troops. Apparently, people in higher authority did not notice the difference between a poorly performing outfit and a poorly led one. This probably had to do with the fact that at the time the Army supported the notion that white southern leaders were the best for Negro troops because the southerners had so much experience giving orders to blacks.

The Korean War would be the first time American troops were truly integrated. General Smith had approximately 1,000 Negro troops. He spread these Marines throughout the 1st Marine Division and ordered officers to be sure that the men knew you were not black or white, you were a Marine.

In 1946, Almond was transferred to Tokyo and became chief of personnel at General MacArthur's headquarters and in January 1949, he took over as Chief of Staff. While Almond was a successful staff officer, his performance in combat would prove to be disastrous at least as far as the Marines and the Army troops in X Corps were concerned.

On September 6, MacArthur gave the order that he wanted the Marines from the Pusan Perimeter to make the landing at Inchon. In one month of fighting, the 5th Marines had suffered 900 casualties but they had helped hold Pusan. After a week's rest and replacing their combat losses, Murray's 5th Marines were shipped north to hook up with Puller's 1st Marines for the Inchon landing. Puller's regiment had arrived in Kobe, Japan on September 2. Marine Air Group 33 which had been at Pusan with General Craig's men would rejoin the 1st Marine Aircraft Wing which was on their way from the U. S.

On September 8, Bowser delivered some more bad news to Smith. "Sir, the Secretary of the Navy has alerted us that parents have complained that their underage sons are being sent to Korea. We are ordered that no Marine that is under the age of eighteen will be allowed to sail."

"How many Marines are we talking about?" asked Smith.

"It will take us a couple of days to look through all the records. When I find out, I will give you a count. One other thing, sir. Have you seen the newspaper this morning?"

"Not yet. Why?" responded Smith.

"You won't believe this, but on September 5, President Truman responded to a congressman who had written him suggesting that the Marine Commandant be accorded a seat on the Joint Chiefs of Staff. According to the newspaper, the president apparently exploded replying, "For your information, the Marine Corps is the Navy's police force and as long as I am president that is what it will remain! They have a propaganda machine that is almost equal to Stalin's.""

"Al, I've heard that the President's dislike of the Marine Corps goes back to World War I when Truman was a captain in the Army Artillery. There was a constant rivalry between the Army and the Marines. He felt that the Marine Corps always got all the glory," explained General Smith. "Probably correctly," he added with a smile.

"So here we are about to make this risky landing at Inchon that an Army general thought up and our president has just stated that we're nothing more than a police force," said Bowser.

Smith said with as much sarcasm as he could muster, "It's nice to hear just how much we are appreciated. We won't mention to the Marines in the landing crafts that they are really MPs."

Two days later, Bowser told Smith that 540 underaged men were not going to Korea. This was a serious loss of man power for a force that was already undersized.

For the next five days Smith and his commanders worked furiously to prepare for the Inchon landing. This would be the first major amphibious assault by U.S. troops since Okinawa in World War II.

The 1st and the 5th Marines of the 1st Marine Division landed at Inchon late in the day on September 15. Rough weather and high seas left many men looking green as they rode in the assault crafts toward Inchon Harbor. Some of the landing craft got stuck in the deep mud flats where they remained until the next high tide. Many of the young Marines were raw and had never participated in any real military action, much less in an amphibious landing. They had little time to prepare and no real practice. But despite all this, their performance in the landing was outstanding. No other major attack from the sea had such an immediate impact on the outcome of a war. Because of the severe thirty-foot tidal changes, the largest anywhere in the world, the Marines' amphibious assault totally took the enemy by surprise. By midnight there were 13,000 Marines ashore at Inchon. The Marines had lost twenty-two men who were killed and 174 who were wounded. By any standard, MacArthur's daring landing had been an unbelievable success.

The next day, Puller's 1st Marines again hooked up with Murray's 5th Marines and fought their way toward the capital of Seoul along the Inchon-Seoul highway.

Smith moved his division headquarters ashore. The Inchon landing had caught the North Koreans off guard. They had few troops in the Inchon area. The battle for Seoul would not have the same element of surprise and the North Koreans immediately began sending reinforcements to Seoul. Intelligence estimates were that there were 20,000 North Korean troops being sent to reinforce Seoul. Smith and Craig set up their living quarters in a ramshackle old barracks that the Japanese had constructed during their occupation. There were very few amenities here; it still had a dirt floor. Smith was Spartan in his habits and never demanded the luxuries that many officers of his rank often expected.

On September 17, General MacArthur stopped by Smith's CP and told Smith he wanted to visit the front lines. MacArthur, along with his entourage of staff officers and select correspondents, made their way to the forward positions. Accompanying MacArthur were Marine Generals Smith and Shepherd, Army General Almond and Navy Vice-Admiral Struble.

Bringing up the rear of the entourage, Shepherd leaned over and said to Smith, "One well-placed North Korean artillery shell could wreak havoc with U.S. military leadership."

Smith nodded and replied, "Thank God they haven't been too accurate so far."

The generals and admiral visited Puller's 1st Marines and Murray's 5th Marine positions. General MacArthur was so impressed by the Marines rapid advance that he presented both Marine regimental commanders with the Silver Star.

General Smith took Puller and Murray aside and individually and privately congratulated them. "You men are doing a great job and are a credit to the Corps. When I first got to Japan, General MacArthur told me that success at Inchon could help save the Marine Corp. So I want to say thanks for helping us all keep our jobs," said Smith in his dry understated humor.

After spending most of the day at the front, Smith returned to his headquarters while MacArthur, Almond, Shepherd and Struble went back aboard the command ship.

When Smith entered his sleeping quarters, he was surprised to see a sergeant setting up a cot next to his.

"Sergeant, what's the extra cot for?" asked Smith.

"Sir, Colonel Bowser asked me to set it up. Apparently, there is an Army general who will be spending the night."

"Sergeant, tell Colonel Bowser that I want to see him."

Five minutes later, Colonel Bowser reported to General Smith.

"Al, who is this Army general that we're receiving tonight?" asked Smith.

"General," replied Bowser, "we just received a call from X Corps that Major General Frank Lowe will be visiting. Apparently, he is a personal representative of President Truman. That's all I know. I figured he should bunk with you."

"That's fine, Al," Smith responded.

A few minutes later as Generals Smith and Craig were about to sit down to dinner, Major General Lowe walked through the door escorted by Colonel Bowser.

"General Smith, General Craig, this is General Lowe," said Bowser by way of introduction.

The three men shook hands and Smith invited Lowe to join them for dinner. Colonel Bowser left the generals and Smith showed Lowe where to put his gear.

As the three men sat down to eat, Lowe said, "Gentlemen, let's dispose of the formalities. Please call me Frank and I will call you O.P. and Eddie."

"Fine with me," replied Smith. "Frank, what brings you to Korea?"

"President Truman has asked me to come over as his personal representative," replied Lowe. "He wants me to report directly to him about what's going on here. The President and I served together in the artillery in World War I. Later, I was then appointed to the Senate War Investigating Committee which then Senator Truman chaired. I was the executive officer assigned to the committee so I worked pretty closely with Harry Truman."

"I remember the committee," said Smith. "At first the military was fearful that it would mess up the war plan by fouling up supply chains, but we all ended up impressed by your success in finding profiteering and stopping it. You helped the war effort."

"I know we did and I have always been proud of my work on the committee," said Lowe. "President Truman and I have been friends ever since. O.P., in effect, President Truman wants me to be his 'eyes and

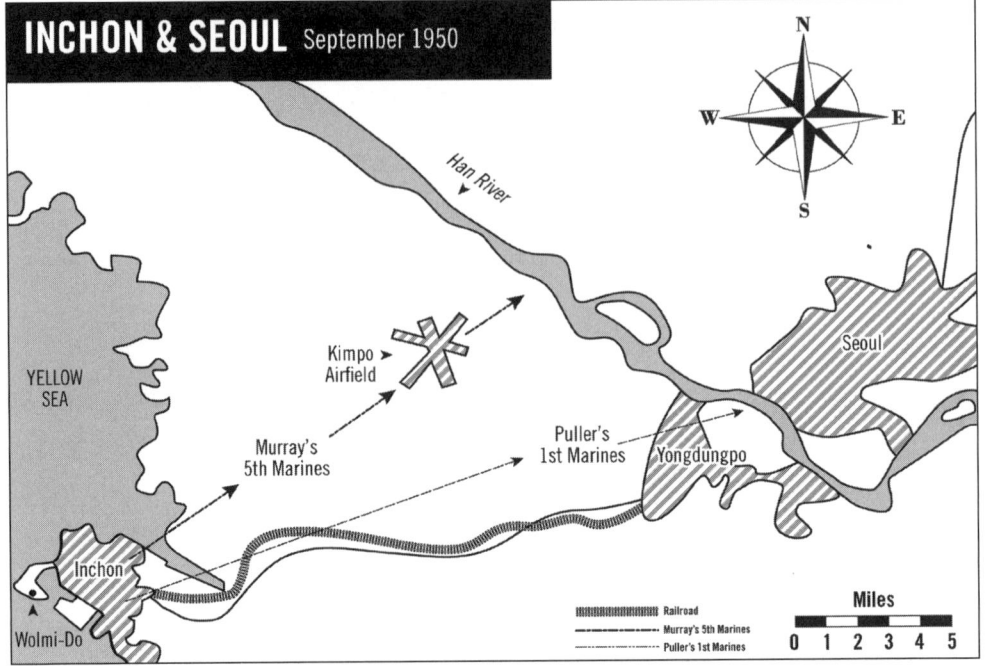

INCHON & SEOUL September 1950

N
W E
S

Han River

Kimpo
Airfield

YELLOW
SEA

Seoul

Murray's
5th Marines

Puller's
1st Marines Yongdungpo

Inchon

Wolmi-Do

Railroad
Murray's 5th Marines
Puller's 1st Marines

Miles
0 1 2 3 4 5

ears' in Korea. Look, I'm not a spy but the President wants to know exactly what's going on. So, with your kind permission, I'd like to visit with your regimental commanders."

"Frank, I'm okay with that. I think you will be impressed," said Smith.

Major General Frank Lowe was sixty-five years old when President Truman brought him out of retirement, but no one thought of him as an old man. Marines of all ranks quickly grew to like and respect General Lowe because he was usually up front where the fighting was and he soon became a familiar figure in his signature pith helmet to the men of the 1st Marine Division. Smith felt comfortable with this man and as he later found out, Lowe always gave the Marines high marks.

On September 18, Murray's 5th Marines took the Kimpo airfield. This was the best air field in Korea and critical to MacArthur's plans. From Kimpo came the close air support for the American troops attacking Seoul. On September 20, Murray's troops started to cross the Han River, the gateway to Seoul, the South Korean capital. Word had spread that

The Savior

they would be facing Russian T-34 tanks that the U.S. rocket launchers couldn't stop. But the Marines soon found their weapons were indeed effective and the North Koreans left a trail of demolished Soviet tanks as they fled north.

Confident that the Marines were rapidly advancing to Seoul and that the North Korean forces were falling apart to the south, MacArthur had a great feeling of optimism. He also still felt the glow of success at Inchon a week earlier. Because of this, on the morning of September 21, he decided to return to Tokyo. Smith accompanied MacArthur to Kimpo Airfield and before the Supreme Commander boarded the plane, he turned to the Marine general and said, "General Smith, it is with great pride and pleasure that I award you the Silver Star. You are the gallant commander of a gallant division." After pinning the medal on Smith's shirt, General MacArthur and his party boarded the plane for the return trip to Tokyo.

Smith looked at the Silver Star and shook his head when Craig offered his congratulations. "Eddie, we both know this award is for valor in battle," said Smith. "It belongs to our men out there who are facing bullets, not to me."

On that same day, Puller's 1st Marines attacked Yongdung-po, a suburb of Seoul, and Litzenberg's 7th Marines, travelling from Camp Pendleton, California, landed at Inchon. Finally, Smith had all three of his infantry regiments in Korea.

The general went out to see Litzenberg as his men were disembarking and heading for camp.

"Good to have you here, Litz," said Smith. "MacArthur thinks we have the war wrapped up but I think your boys will be seeing plenty of action. We still have to take Seoul and then I imagine we'll have some chasing of the North Koreans. These Marines look pretty young, Litz. You did get the word about recruits under age eighteen, didn't you?'

"Sure, General. That order prevented several of my Marines from sailing with us," responded Litzenberg. "Many of these men are green, but we've been working hard with them at Pendleton and on the boat here. They will be ready when they are called on."

The battle for Seoul caused another rift in the already tenuous Almond-Smith relationship. MacArthur wanted the city retaken by September 25, three months to the day from the start of the Korean War. And General Almond, ever responsive to MacArthur's wishes, pushed Smith to secure the city by this artificial date.

Almond told Smith on August 23 that the slow pace of the Marines was not acceptable and that if they didn't gain ground in the next twenty-four hours that the Army's 7th division infantry would be sent in to do the job. This rankled Smith.

On the twenty-fourth, Almond gave orders directly to Puller and Murray without first going through their commanding officer, General Smith. When Smith found out about this, he was furious and for the first time actually confronted Almond. "I would appreciate it if you would not give orders to my regimental commanders; you give the orders to me and I will make every effort to see that they are carried out promptly."

Almond offered no apology for this breach of command and denied having done this, simply stating, "There must be some misunderstanding."

Smith bit his tongue and just walked away. He had spent most of his life in the military and knew there was little future in getting into a pissing contest with a man who outranked him. But that very night, another incident would occur where Smith would question Almond's judgment and his ability to command.

There were reports of the enemy fleeing the city of Seoul but these had not been confirmed by Smith's intelligence people. At 8:00 pm, Almond sent orders for the 1st Marine Division to attack. Smith's Marines were exhausted after nine days of continuous fighting. They had only been able to sleep a couple of hours at a time. Launching an attack in the dark in an unfamiliar, Asian city of a million people made absolutely no sense. The enemy mined roads and were fighting from barricade to barricade to retain the capital city. Smith immediately questioned the order saying, "I think it is inadvisable to wage a night attack, especially since we don't think the enemy is actually withdrawing."

In what was becoming a typical exchange between Almond and Smith, X Corps replied that the attack must proceed without delay. Smith's on the scene assessment was once again disregarded.

On hearing the X Corps order, Craig immediately sought out Smith and asked, "What the hell are these guys thinking? How is waiting ten hours to attack going to change anything here?"

"I don't know," said Smith. "MacArthur dictates and Almond doesn't seem to have the sense of a turnip. Sorry Eddie, forget I said that. Sometimes I wonder how we won the World Wars. We could have encircled Seoul and taken it without destroying the city and putting so many civilians at risk. But instead we are following the MacArthur/Almond plan of flattening the city with air strikes and finishing it off with heavy artillery."

Smith did manage to delay the attack until 2:00 am. However, the North Koreans attacked first, thereby preventing the Marines from walking straight into a disaster. The battle ended early the next morning with the North Koreans suffering heavy casualties. Smith knew that Almond's foolish order had surely put his division in jeopardy. From now on, Smith would question the wisdom of every Almond directive.

Late on the night of the twenty-fifth, Almond prematurely announced the liberation of Seoul to the press with MacArthur echoing these words the next day in Tokyo. The city was far from liberated, however, as Smith and the Marines in the city knew. American artillery had badly damaged Seoul and the North Koreans used the rubble for street barricades. After three days of intense fighting from September 25 to the 28, the Marines finally secured the city of Seoul. During this time, the Marines suffered 711 battle casualties after the Army hierarchy claimed the city was liberated.

Finally, on September 29, in a ceremony filled with pomp and circumstance, General MacArthur returned the city of Seoul to the

United States-backed Korean leader, Syngman Rhee. What he gave back was a sorry mess. The city had been occupied by the North Koreans just weeks before and now had been a battle ground again to drive the North Koreans out. Bombed out buildings and roads provided poor shelter to the hundreds of thousands of refugees who had fled to the city for safety. People were sleeping in the streets. The city reeked from the burned out wreckage of buildings and the swollen, decaying bodies, as well as the swell of humanity sheltering there. Against this backdrop, MacArthur in flowery oratory, returned the capital to the South Koreans.

Unfortunately, there were only six Marines attending the ceremony in the war-torn government palace: Generals Smith and Craig, Colonels Puller and Murray and two aides. They were still dressed in their dirty combat clothes.

When the ceremony was over, the six Marines walked out of the palace. They were furious and rightfully so.

"I feel like somebody's poor relation," said Puller.

He looked at General Smith and said, "The Marines did all the fighting, we liberated the damn city and the Army gets all the honor. MacArthur brings in all these spit shined Army military police from Tokyo and he lines the palace with these guys. They never fired a shot! There wasn't a damn word said about the Marines. Who in the hell did they think did all the fighting?"

General Smith, in a voice filled with disgust said, "Lewie, that's just typical MacArthur showmanship. He likely thinks we should be grateful for being invited today."

Smith was justifiably proud of his Marines, both at Inchon and in the battle to take Seoul.

"I think our men do us proud," he told Craig as they left the MacArthur ceremony at the palace. "Even if they are not getting the credit they deserve."

The Almond-Smith, Army-Marine relationship was further tested by the appearance of General Smith's picture on the cover of *TIME* magazine on September 25. Unlike Almond, Smith never sought publicity and was unaware of the cover story. He usually went out of his way to avoid the press. The article discussed the Korean War and the Inchon landing and praised Smith and his 1st Marine Division. It credited Smith as the expert on amphibious landings while Almond's name was barely mentioned in the *TIME* article. The focus was clearly on Smith and his background. This only further aggravated the relationship between Almond and Smith. Smith's only reaction to the article, once he had seen it, was to comment to Craig that he thought at least it was good publicity for the Marines who had borne the brunt of both the Inchon and Seoul actions. He also was pleased to know how proud his wife, Esther, and his two daughters were to see his face on the newsstands.

General Cates, Commandant of the Marine Corps and Smith's old boss, arrived in Korea on October 2 to review the 1st Marine Division.

"You have a lot of Marines smiling back in Washington," Cates told Smith.

Smith looked puzzled, but Cates continued, "The *TIME* magazine article, O.P. It made a lot of your old Washington friends very pleased."

After a briefing at Smith's headquarters in Seoul, the Commandant met with the regimental commanders and Smith presented the Commandant with the North Korean flag that was taken down from the government palace in Seoul. Cates, at least, knew what the Marines were doing in Korea. It was a proud moment for Smith.

As Cates was leaving he said to Smith, "With the North Koreans trapped between the Marines in the north and the 8th Army in the south, the Korean War should be just about over, don't you think, O.P.?"

"Well, MacArthur's telling us we will be home by Christmas," responded Smith. "So, we can hope that just wrapping up here will be the end of it, but somehow I don't really believe that's how this is going to play out."

5

A WARNING IGNORED

Just days after Seoul was liberated, the slumbering giant that was China issued a warning. Shortly after midnight on October 3, Zhou Enlai, Communist China's foreign minister, told the Indian Ambassador, K.M. Panikkar, that if U.S. Troops invaded North Korea, China would enter the war. If only South Korean soldiers crossed the border, China would stay out. Panikkar delivered the message to the U.N. This information was passed on to Congress, the President and to MacArthur's headquarters in Tokyo, but the Supreme Commander of the Allied Powers ignored the warning.

On October 15, General MacArthur met with President Truman on Wake Island. Truman came to the Pacific because MacArthur claimed he couldn't possibly leave the war to meet him in Washington. Truman wanted to know if MacArthur believed that if the U.S. invaded North Korea that the Chinese would enter the war. The General informed the President that it was unlikely and MacArthur said he thought that North Korea's resistance would end by Thanksgiving. By a strange coincidence, the Chinese picked the same date, October 15, to begin moving their forces across the Yalu River, which formed the border between China and North Korea, into the mountains of North Korea.

MacArthur shared his plan for invading North Korea with his generals. He would divide his army; the 8th Army, under the command

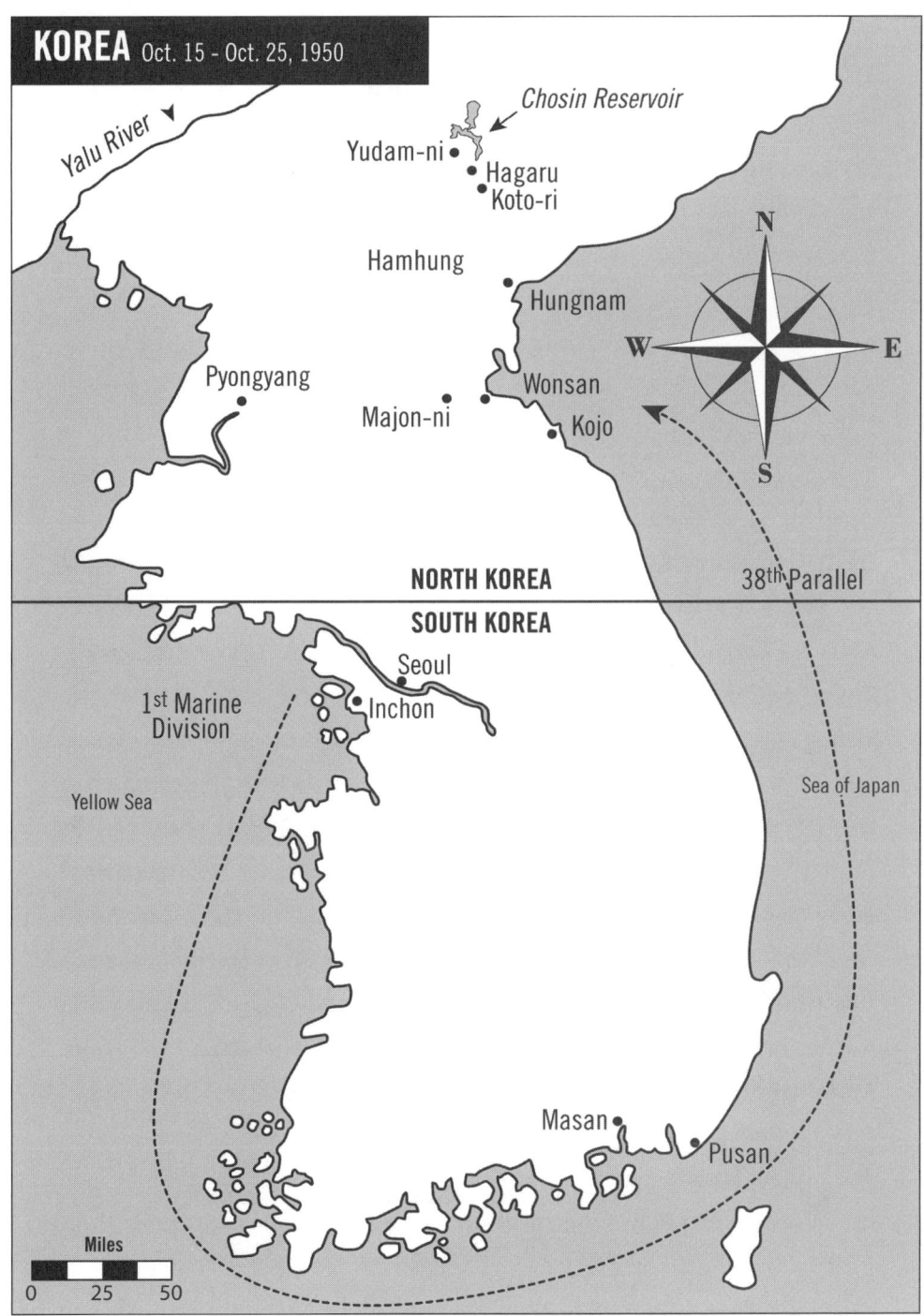

KOREA Oct. 15 - Oct. 25, 1950

Yalu River

Chosin Reservoir

Yudam-ni

Hagaru
Koto-ri

Hamhung

Hungnam

N

W E

Pyongyang

Wonsan

Majon-ni

Kojo

S

NORTH KOREA

SOUTH KOREA

38th Parallel

Seoul

1st Marine
Division

Inchon

Yellow Sea

Sea of Japan

Masan

Pusan

Miles

0   25   50

of General Walton Walker, would go up the northwest coast of North Korea and X Corps, made up of the 1st Marine Division and parts of the U.S. Army's 7th Infantry Division, would go up the northeast coast. Both would push on to the Yalu River on the Chinese border. Douglas MacArthur had no plans to merely restore South Korea to its status of four months earlier. He had rolled the dice and had won big at Inchon. He had retaken Seoul and had the North Koreans in retreat. Now there was one more feather to put in his cap: a united, non-communist Korea.

Instead of packing up his multinational force and sending them to their home countries, MacArthur's ego pushed him north. MacArthur's landing at Inchon turned a crushing defeat into a glorious victory. But when he decided to invade North Korea, in one of the most ironic turn of events in the history of war, he turned a stunning victory into a humiliating defeat.

While MacArthur was meeting with President Truman, Smith's 1st Marine Division loaded onto transport ships and left Inchon on October 15 just one month after their invasion of the port. Four days later after sailing completely around South Korea through stormy seas, they arrived at the port of Wonsan in northeast Korea. They were 110 miles north of the 38th parallel, deep in North Korea. Because the harbor was heavily mined, the ships turned around and headed south. They reversed course every twelve hours heading north and then south and then starting all over again while the harbor was being cleared by Navy frogmen. During this time food supplies ran low and dysentery broke out. Fresh water was rationed. It was a miserable time for the Marines aboard the ships. Many would sleep on the decks rather than their stuffy quarters. Finally on October 26, the harbor was cleared and the Marines went ashore. There were no North Korean forces to oppose them in Wonsan. In fact, they learned that Bob Hope and his USO tour had put on their show for Marine aviators at Wonsan on October 24, and had flown out by the time the Marines landed. Still they set up a perimeter, dug foxholes and prepared for the next leg of their journey. The race to the Yalu was on!

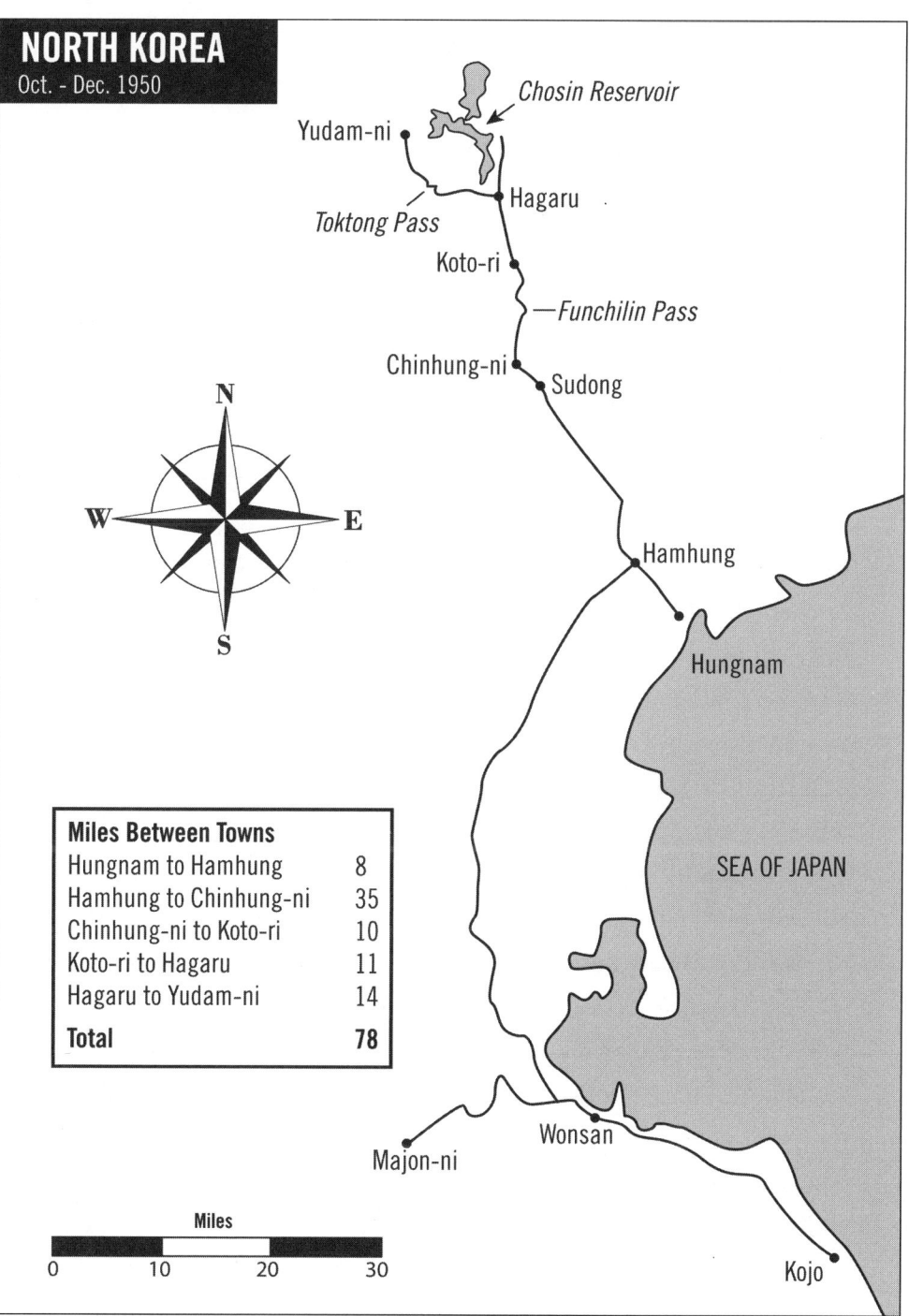

NORTH KOREA
Oct. - Dec. 1950

Chosin Reservoir

Yudam-ni

Hagaru

Toktong Pass

Koto-ri

Funchilin Pass

Chinhung-ni   Sudong

N

W   E

S

Hamhung

Hungnam

SEA OF JAPAN

Miles Between Towns

Hungnam to Hamhung	8
Hamhung to Chinhung-ni	35
Chinhung-ni to Koto-ri	10
Koto-ri to Hagaru	11
Hagaru to Yudam-ni	14
Total	**78**

Majon-ni

Wonsan

Kojo

Miles

0   10   20   30

6

THE DRIVE NORTH

Soon after landing at Wonsan, the Marines were issued their cold weather gear from supply. The men were happy to have this as the heat and humidity that had met them in Korea was changing and the weather was already getting quite chilly. The Marine Corps used the "layered principle" for winter clothing. Each Marine was given three sets of woolen underwear. Wool trousers and a flannel shirt were worn over the underwear. A heavy sweater and a pair of waterproof trousers would form a third layer. On top of this was a field jacket. The final layer was a bulky, pile lined parka with a hood.

Marines were also issued three pairs of heavy wool socks, two pairs of felt innersoles, and a pair of shoe-pacs (rubber bottomed boots with lace-up leather tops). Mittens were supplied with wool inserts with a trigger finger opening. Finally, each Marine was given a mountain sleeping bag, equipped with a quick release zipper to allow speedy extraction of the occupant.

The winter gear added more pounds to the already heavy load each Marine carried. The men groused about this added burden but in the coming weeks would find that it saved their lives. The many layers of clothing would also decrease their mobility, but again the men would tolerate the additional bulk for the warmth it provided.

After landing at Wonsan the Marines began their trip north. Truck convoys clogged the narrow roads to a place they had never heard of:

the Chosin Reservoir. The Koreans called it Changjin Reservoir, but the United States used the name given to it by the Japanese during their occupation because the maps MacArthur had in Tokyo used the Japanese name. And Chosin rhymed with frozen and so the name stuck.

Many of the roads in North Korea were unpaved, dirt roads full of peasants in ox carts. The truck convoys on these roads raised dust clouds that choked everyone. Vehicles had to be coordinated by radio to avoid traffic jams. But except for some bunkers and other evidence that the enemy had been in the area, the troops did not meet resistance for most of the journey north.

They traveled through a bleak, damp countryside. Terraced brown rice fields bordered the road. Poplars and pine trees covered the surrounding hillsides. Communist Russian posters were tacked up in every village they passed through. The traveling Marines saw many local Koreans on the road collecting any items discarded by troops that had already passed through. These Marines needed to stay constantly alert because there was no way to know if these locals were friend or foe.

While Smith and his division command were planning for the attack in the north, General Almond was also busy doing some reorganization. Almond, without consulting Smith, split up the 1st Marine Division. The three fighting arms of Smith's forces were sent to separate locations. Litzenberg's 7th Marines were sent north to the Chosin Reservoir; Murray's 5th Marines would follow; Puller's 1st Marines would be separated into isolated battalions and sent south to Kojo and west to Majon-ni.

General Craig brought Smith the news.

"Eddie, I don't want to sound paranoid, but do you know any reason why Almond could think this is a good plan?" asked Smith.

Craig looked at Smith and responded, "No, General. I don't see the logic of splitting up the forces either. But I don't see how we can avoid complying with these orders unless we want to be shipped home in disgrace."

"I know," said Smith. "I would try and change Almond's mind on this, but have had no success with that in the past. But we have to keep on top of this and make sure our men are well supplied and not too strung out.

"Sometimes I doubt myself," Smith told Craig. "I opposed the Inchon landing and it turned out amazingly well. Maybe MacArthur is a brilliant strategist and his plan will do the job. He is calling this an 'end the war offense.'"

On the evening of November 1, Marine Lieutenant General Lem Shepherd visited and had dinner with O.P. Smith at Wonsan on the North Korean seacoast. While Shepherd had been passed over for commander of X Corps, he still was in Smith's chain of command.

Smith had a lot on his mind and so after dinner the two Marines talked frankly.

"First," said Smith, "I wanted you to know how disappointed we all were when we heard you were not going to command X Corps. Life would be better if you were in charge."

"I appreciate the sentiment," responded Shepherd, "but we both know how little control we have over these things. Politics and commands are always moving targets."

Smith lit his pipe and said, "Lem, I've got a lot of concerns about this operation. This 'race to the Yalu' is misguided."

"Go on," urged Shepherd.

"Well," continued Smith, "Almond has all my regiments spread out all over the place. If you concentrate these three regiments, you have a powerful fighting force, but when you divide them, you weaken them. The 8th Army is eighty miles to our west and we're separated by the Taebaek Mountains. Our left flank is wide open and if any of these outfits gets in trouble, we can't support each other. And then there is this road leading up the mountains. From Hungnam all the way north to Yudam-ni where Murray and Litzenberg are headed, the road is seventy-eight miles long. From Chinhung-ni to Yudam-ni, it is one lane, dirt and gravel. It's the only road and this will be our main supply route. What if the North Koreans or the Chinese close it? Or the weather makes it impassable? What happens then? We are cut off!"

"O.P., as far as we know the few Chinese we've come across are just volunteers. There is no evidence that the Chinese plan to make good on their threat from last month," countered Shepherd.

"I'm not so sure about that, General," said Smith. "But if trouble starts then these Marines are going to have to live and fight in these mountains in the bitter cold when winter comes. Attila the Hun wouldn't want to fight up here. This whole thing is misguided and Almond is almost reckless in charging up into the mountains at this hurried pace. MacArthur says 'jump' and Almond asks 'how high.'"

After lighting a cigarette, Shepherd said, "O.P. play the game, don't get so mad with Almond. He is trying to do the right thing. You've got the North Koreans on the run. You've got the momentum. When the enemy is retreating, you've got to go after them."

Smith just shook his head. He was surprised and disappointed to find that Shepherd was unsupportive and unsympathetic to his concerns. It seemed to Smith that Almond and now Shepherd looked at this as merely a "mopping up" operation that would be finished by Christmas, not as the major offensive that Smith saw when he looked at the maps on his desk.

They talked some more and Shepherd left shortly before midnight.

That night Smith couldn't sleep. Ominous thoughts kept nagging him. He wondered why Shepherd wasn't agreeing with him. The man was an experienced combat veteran whom Smith respected. Was it because Shepherd was in line to be the next Commandant of the Marine Corps and didn't want to rock the boat with Almond or MacArthur? Was he worried about looking like a sore loser after not getting the X Corps command? Were politics and self-promotion putting the Marines' lives in jeopardy?

Smith wondered if Shepherd was recalling what happened to Holland Smith. In World War II, during the Battle for Saipan, the commanding Marine general, Holland M. Smith, had relieved an Army general, Ralph Smith of his command and it created a lot of animosity between the Marine Corps and the Army. As a result of this, Holland Smith was not given command of the 10th Army for the invasion of Okinawa. Instead, he was moved out of the combat zone. Maybe Shepherd didn't want any friction with the Army that would jeopardize his chance of becoming Commandant. He didn't want to be in the middle of any Army-Marine controversy.

Or was it because Shepherd and Almond had both graduated from the Virginia Military Institute; Smith knew that Shepherd had graduated in 1917 and found out that Almond had graduated in 1915. They clearly overlapped in the time at school. Maybe Almond and Shepherd had been friends or were still friends.

Whatever the reason, Smith was on his own, and he now knew it. He wouldn't be getting any support from the Marine above him in his chain of command and Almond had made it clear that he thought Smith was "difficult" and full of excuses for not performing the orders Almond issued in the required time frame.

Smith's style in commanding a division was molded by his experience as the assistant division commander of the 1st Marine Division during the battle of Peleliu. Before this battle started the commanding general of the 1st Marine Division, General Rupertus broke his ankle, hobbling him for most of the fighting. Prior to the attack, Rupertus told Smith and his three regimental commanders, "There will be no change in the orders, regardless. Even if General Smith attempts to change my plans or orders, you regimental commanders will refuse to obey." Smith knew that this was certainly a poor use of an assistant division commander and determined that it was not an example he would ever follow. He had Rupertus' job now with the 1st Marine Division and took care to make sure he used his staff wisely.

Peleliu was one of the bloodiest battles of World War II. It was also a massive intelligence failure. The island was reported to be flat but it wasn't. Running down the center of the island was a 556-foot mountain. It was an underground fortress filled with caves and bunkers that the Japanese refused to surrender.

Rupertus had told the press that the fighting would be rough but would be over in a few days. Instead, the battle dragged on for weeks and the 1st Marine Division suffered over 6,500 casualties including 1,252 killed. His infantry regiments had an unbelievable 50% casualty rate. But Rupertus kept pushing his regimental commanders to get the battle over, which only increased the number of casualties. Rupertus' prediction of a short but rough operation had turned into a bloody nightmare. Smith had been there to see it all.

After the battle was over, Rupertus returned to the States where, despite his abysmal combat record, he was named Commandant of Marine Corps Schools in Quantico. His foolish combat strategy on Peleliu would never be explained as he died of an apparent heart attack a few months later.

But Smith had lived through this inept leadership. Peleliu had taught him four valuable lessons: the need to be cautious, to pay attention to the intelligence, to understand the terrain and to never underestimate the enemy. These lessons learned would serve Smith and his 1st Marine Division well in the weeks ahead. He only hoped that MacArthur's prediction of a speedy victory would not end up like Rupertus' similar declaration to the press.

Smith knew that the tactics of fighting in North Korea would again be largely dictated by the terrain. The road leading north either was cut into the side of a mountain or wound its way through valleys. The shoulders of this road had to be protected before the Marines could advance. This would be the accepted principle going forward: they would take and hold the ground on both sides before advancing. Everyone knew that there was only one way in and one way out.

On the morning of November 3, while Smith and Craig were having breakfast at their headquarters in Wonsan, Colonel Bowser came in.

"General Smith, General Craig, good morning sirs," he greeted.

"Good morning Al, have a seat. Want some coffee? What's up?" inquired Smith.

"As you know Colonel Litzenburg and his 7th Marines are headed north on the main road to Chosin Reservoir. They are about twenty miles out of Hamhung at a place called Sudong.

"Sir, Litzenberg called in to report that they got hit hard last night at Sudong," said Bowser.

"How many casualties?" asked Craig.

"Sir, Colonel Litzenberg doesn't have any numbers yet."

General Smith took a sip of his coffee and asked, "Who attacked them? Do we know? The reports from yesterday said they were halfway to Chosin Reservoir and meeting only scattered resistance."

"Litzenberg said he thinks that they were hit by a Chinese division. They saw some enemy tanks," answered Bowser. "But they held their ground all night."

Smith looked at Craig and said, "I want you to fly up there Eddie and find out what's going on. I don't like the sound of this."

Bowser interjected, "Certainly doesn't sound good sir. I'm afraid there's more bad news."

"What's that Al?" asked General Smith.

"Sir, the 8th Calvary, which is part of the Army's 1st Calvary Division got hammered yesterday at a town called Unsan up on the northwest side."

"Who attacked them?" asked Craig.

"Preliminary reports say it's the Chinese again, General."

"Al, get all the information you can about what happened to the 8th Calvary," said General Smith. "Gentlemen," he continued, "we've got to make sense of this. We've got Chinese at Sudong, we've got Chinese at Unsan. I'm afraid there's something big going on and we need to know just how big it really is and fast. Eddie, get a helicopter ready and fly up to see Colonel Litzenberg right away. General Almond called yesterday and wants us to go after guerillas with our patrols. It sounds like we have a lot more to deal with here than a few guerillas."

Colonel Bowser took a last sip of his coffee and went back to the command post.

General Craig called to order a helicopter, quickly finished breakfast and then went to the waiting copter for his trip north.

General Smith walked to his command post puffing his pipe to continue planning the move of his headquarters north from Wonsan to Hungnam. They say an army travels on its stomach, he thought, but it seems like we travel with a mountain of papers. And tomorrow all of this paper is heading north.

Thirty minutes after boarding, Craig's helicopter descended to a landing in the Sudong valley. There he saw hundreds of bodies littering the hillsides and shoulders of the road. The helicopter sat down on the road not far from Litzenberg's command post.

Craig stepped from the helicopter and looked at the steep hills that climbed over 2,000 feet from the floor of the valley. He knew that this rugged terrain would be hard to fight in, especially in winter.

After a few minutes, Litzenberg's jeep pulled up to Craig's helicopter. "Good to see you, General," said the colonel extending a handshake. "Hop in and I'll take you up to my command post."

Craig climbed into the front seat while Litzenberg got into the back. The driver sped away and a few minutes later they arrived at Litzenberg's headquarters.

The jeep stopped and Litzenberg said to Craig, "I want to show you something."

They walked a short distance where they saw about fifty Chinese prisoners sitting on the ground being guarded by Marines with rifles at the ready.

Litzenberg spoke first.

"General, our intelligence people confirm that these are Chinese communist soldiers and I guarantee you they are not just volunteers."

"Why are you so sure of that?" asked General Craig.

"Because they have all said the same thing. They are all part of the 124th Chinese Communist Division. They say they have been on the move since mid-August, and their mission was to block our movement north. We estimate that we've killed over 1,000 of their men. Hard to believe that thousands of Chinese 'volunteers' want to come through these mountains with winter coming on to help their North Korean neighbors. Our reconnaissance also reports finding bunkers and foxholes in the hills. And sir, the bodies of the enemy, like these prisoners, appear to be seasoned soldiers. They aren't raw recruits."

Litzenberg pulled a crumpled pack of Lucky Strike cigarettes from his field jacket. "Damn," he exclaimed, "hard to keep these from breaking." He straightened one of his cigarettes, put it in his mouth and lit it. He inhaled the smoke deeply before blowing it out.

Craig then asked, "What about your casualties?"

Litzenberg reached into the upper pocket of his jacket and took out a small notebook. He opened it saying, "We had forty-four killed, another five died later of wounds, one missing and 162 wounded. Now we know we are in a war."

Both men were silent in the face of these numbers. For these Marines, there really was no tomorrow thought Craig. They spent some time going over tactical and logistical matters before Craig left the command post. He walked back over to where the Chinese men were sitting.

Craig looked at the prisoners for what seemed like a long time then he turned to Colonel Litzenberg and said, "I agree with you, Colonel. These men aren't volunteers. Your men did a good job last night. I saw hundreds of enemy dead out there. As you probably know, this is the biggest battle the Marines have fought in North Korea. I'll report all this back to General Smith. Have your driver take me back to the helicopter. Something is going on here, Litz. Be careful!"

With that, General Craig left in the jeep and then he flew back to Wonsan.

When the helicopter landed in Wonsan, General Craig went immediately to meet with General Smith.

Craig briefed the general on the battle and gave him the casualty counts both for the Marines and the enemy. They discussed the Chinese prisoners that were captured and what Litzenberg's patrols were seeing.

"Litzenberg is certain that he is fighting the Chinese, not North Koreans or some rogue Chinese mercenaries or volunteers," said Craig.

"Yes, no matter what MacArthur and Almond are saying, I trust Litzenberg's assessment. He is on the ground seeing it first hand," said Smith. "His patrols have captured a number of prisoners and they confirm that they are Chinese. His intelligence has to be more reliable than what Tokyo is telling us. We need to be sure Almond knows what's happening. Please be sure to get your report to him."

Late in the morning the following day, Colonel Bowser reported in to General Smith.

"Sir, I have more complete information about what happened to the 8th Calvary over in Unsan. They had over 800 casualties, General. This is a pretty good outfit. These soldiers aren't a bunch of rookies. And they also lost a lot of their equipment, including twelve 105 howitzers, nine tanks, 125 trucks and a dozen recoilless rifles. They report that it was definitely Chinese who attacked them.

"There were also a significant number of 8th Calvary soldiers who were captured. I'm not certain of the numbers yet. The report I got, sir, is that it was a massacre."

"Dear God," said Smith. "This isn't how this offensive should be getting started."

"Al, have you seen Eddie Craig?" asked General Smith.

"He briefed me this morning about what Litz ran into in the Sudong Valley. It's definitely the Chinese driving this war now."

Bowser and all of Smith's staff agreed who the enemy was. But efforts to make the Supreme Commander understand were unsuccessful. After conversations with Tokyo, Bowser reported back that MacArthur, his intelligence chief, General Willoughby and General Almond all chose to disregard the facts.

"As far as they are concerned," said Bowser, "the Chinese that the Marines and soldiers were fighting were all volunteers. This is an incredible intelligence failure and one that MacArthur's staff has refused to acknowledge. Who knows how many thousands will be killed or injured before they understand?"

Craig added, "After these two battles, the Chinese forces disappeared into the snow covered mountains. I looked for them this morning when I flew to Sudong and couldn't see a trace. They remain camouflaged during the day when reconnaissance planes fly overhead. Because Willoughby's people don't see them, they deny their existence. General Lowe has told us that MacArthur assured President Truman on Wake Island last month that the Chinese would not enter the war. He hopes that by failing to admit this fact, it won't be true."

The goal of MacArthur, which was to reach the Yalu River and unify all of Korea, and the goal of Washington, which was to avoid a war with China, were at total odds with each other.

General Smith moved his command post north from Wonsan to Hungnam on November 4, in order to be closer to his men. Smith flew to Hungnam by helicopter while most of his headquarters staff arrived by rail later that evening. The bulk of his headquarters would remain there during the coming operation. His command post occupied an abandoned engineering college on the outskirts of the city. Strangely enough, in one of the rooms there was a mortuary slab. When Smith first saw it, he told his aide, "Get that thing out of here! It's too depressing. There is enough death around us without having that thing in here."

As the men worked to set up the command post, Craig said, "General, I happened to see your photo today."

"Where?" asked Smith.

"Several men have gotten copies of *TIME* magazine with you on the cover. Their families mailed them from home. They are really very proud to see a Marine get some recognition, and proud that it's their commander."

"That's kind of you to say," said Smith. "It might also be the reason my Korean translator was asking me about roses. That reporter had to include that my hobby was gardening and my specialty was roses. I actually worked my way through the University of California working as a gardener and found I liked the work. It might be something to do when I hang up my boots."

Eddie Craig replied, "Sir, we all hope that doesn't happen for a long time. It was a great article and after that insulting fiasco at the palace ceremony in Seoul, it's about time that you and the 1st Marine Division got some recognition. By the way, I hear General Almond has caught up with you, General."

"What?" asked Smith.

"He got his mug on a *TIME* magazine cover too," said Craig. "I haven't seen the article or magazine, but hear it's on the newsstands. Our men will not be holding onto too many copies of that one. You still got there first, General Smith."

"I just wish we were all on the same page in how to wage this war," responded Smith.

The next morning, on Sunday, November 5, General Smith left the chaos of setting up the new command post to his staff to personally fly up to see Litzenberg. The losses of so many men weighed on him. Litzenberg's command post was in a deep valley twenty-five miles south of Chosin Reservoir at the foot of the mountains.

Litzenberg met the helicopter and took Smith to the command post.

"I ordered air strikes this morning," said Litzenberg. "The noise is music to my ears."

Smith agreed. "Our air support is the key advantage we seem to have right now. Intelligence is not finding any air power on the other side, but given how little we really know about the enemy, we can't assume they don't have some air support or that some isn't on the way from China or Russia."

The two Marines discussed the fight ahead and General Smith then returned to Hungnam.

The next morning, Smith told his staff about the Litzenberg meeting and that they would soon be heading into the mountains.

Later in the afternoon, General Frank Lowe arrived at Smith's command post. President Truman's eyes and ears in Korea was a busy man. The usual Army-Marine animosity had never been a part of the Lowe-Smith relationship. They had developed a sincere mutual respect and friendship. Lowe told Truman that Smith was not concerned with acquiring real estate but with finding and killing the enemy with as few Marine casualties as possible.

Lowe had just been with the 8th Army and had seen the horrific losses inflicted by the Chinese army. Smith thought he seemed very depressed.

"What do you think happened over there, Frank?" asked Smith.

"For starters, the South Korean soldiers took off when the fighting got bad. We're here to win back their country and these soldiers left the whole U.S. Army battalion surrounded with no help and no way to escape. I'm afraid several senior Army officers will get relieved but not if I can help it. In this crazy war, you can end up in hot water for following orders and that worries me. Makes you wonder if you shouldn't use your own good sense."

"I agree," responded Smith. He hoped Lowe would remember this conversation if X Corps tried to get rid of him for doing exactly what Lowe was advocating. In his conversations with Lowe, he found that Lowe respected MacArthur, so Smith was careful about any criticism.

The next day on the seventh of November, General Almond arrived at Smith's command post. Smith wasted no time in confronting Almond.

"In view of what happened at Unsan and also the fight we had at Sudong, I think it's important that we take a good look at what's going on."

Smith could tell that Almond had been stunned by the events of the past week and he pressed him hard to reconsider the drive north.

"What do you have in mind O.P.?" asked Almond.

"Look, winter is coming on early this year and it is going to be very difficult to supply my men up in the mountains. My recommendation is that we hold all the ground around Wonsan, Hamhung and Hungnam and we don't go north of Chinhung-ni. I would also like to bring my regiments together. Parts of my division are as much as 170 miles apart. Puller's 1st Marines are scattered around Wonsan; that's fifty miles south of here. Murray's 5th Marines are heading northeast to Singhung and Litzenberg's 7th Marines are going northwest to Koto-ri. My regiments are being thrown in all different directions. This is a poor and dangerous use of my division. It weakens our fighting ability and plays havoc with supplying these men."

Almond sat silently for a moment, his eyes narrowed as if he were contemplating a painful decision. He replied, "O.P., I will agree to concentrating your division but I want your Marines to hold Hagaru at the foot of Chosin Reservoir. Our orders are to head north and General MacArthur wants this war over by Christmas. We can't just have all our troops sitting here waiting for spring."

Smith did not like this advance even further to the north but at least he got a promise from Almond that he could bring his regiments together.

The Marine Corps' 175th birthday was celebrated on November 10, in General Smith's mess with most of his staff present. Traditionally the birthday is celebrated by all Marines worldwide. As is the custom, Smith read the birthday message from the Marine Corps manual and then cut the cake with a Korean sword. In keeping with tradition, the first piece of cake went to the oldest Marine present, Brigadier General Eddie Craig. The good natured banter about the "old man" just barely concealed the concern of all present about the task ahead. The dead Marines from the previous week put a damper on the festivities. And all present knew that more casualties would follow.

Litzenberg and his 7th Marines were in trucks on the way to Koto-ri. Once they arrived, they would set up their artillery in a tree nursery and begin scrounging firewood. It would be a very cold birthday celebration for these Marines.

The next morning, Litzenburg called Smith on the radio, "Sir, last night the temperature dropped from forty above to eight below zero and the wind was blowing hard. My Marines were just stunned by the cold and over 200 of my men went into shock."

"How serious is this?" asked Smith.

"Sir, this weather came out of nowhere. The doc says after getting them warmed up and a day's rest, they will be returned to duty. I wanted to give you a heads-up on this. The cold up here is almost unbearable and it's still early November."

"Thanks Litz. I'll pass it on to Colonels Puller and Murray," said Smith. "This is going to be a huge problem and I just need to make sure we are prepared for it. Keep me informed on how your men are recovering."

On November 11, Smith received orders from Almond that the objective for the 1st Marine Division was once again a rapid advance to the Yalu River which formed the boundary between China and North Korea. Smith was again dismayed. He called in Craig and Bowser to review Almond's plan.

Smith looked at both men shaking his head, "Here we go again. Now Almond wants to push north to the Yalu and fast. Two days ago, he told me we had to hold Hagaru at the foot of Chosin Reservoir. Now we are supposed to drive north again. Apparently X Corps and MacArthur's command in Tokyo can't face reality. They discount the

Chinese we have captured and they call them volunteers. How ridiculous is it to believe that a soldier from a communist country like China would be a volunteer? It makes no sense! My instincts and all the intelligence we have gathered tell me that the Chinese are in this in a big way, no matter what MacArthur and his cronies think. I know MacArthur hit a home run at Inchon, but he's not here seeing what we're seeing. Frankly gentlemen, I'm very concerned over the lack of realism in Almond's plans. He wants a rapid advance to the Yalu River but he is ignoring the Chinese capabilities not to mention the weather. My God, they might have a million soldiers up there!"

"General," said Bowser, "it would be foolhardy of us to push further and faster, spreading ourselves thinner and thinner, not until we know what we're up against. Hell sir, it already feels like we're being pushed out at the end of a long, snow covered limb."

Smith added, "It's about 150 miles from Chinhung-ni to the Chinese border and Almond wants us to advance north along this terrible road with winter coming on. And then there's our left flank. It's wide open! Look at this map. The nearest American unit is the 8th Army and they are ninety miles to the southwest. They have recovered from the attack at Unsan but from all reports I'm hearing, the new Army troops are not well trained. Between us and the 8th Army are the Taeback Mountains. We have no idea how many Chinese might be hiding in there and we know that the Korean road system is something from the dark ages. The 8th Army can't help us and we can't help them."

During Smith's thirty-three year career in the Marine Corps, he had faithfully obeyed every order that he had been given, whether he liked it or not. And there had been many orders he didn't like. But this order to rush to the Yalu was different. It made absolutely no sense. And now Smith was responsible for thousands of lives. The stakes were just too high to be wrong. Maybe he had been wrong about Inchon, but all his experience and instincts told him that following this order could be disastrous.

Smith slowly relit his pipe, tossing his spent match into an empty ration can sitting on his desk. Smith was thinking of what to do. Finally, he said, "We have to use some discretion here." Smith knew he could get relieved for disobeying orders, but he said, "Eddie, tell Litzenberg and Murray to proceed slowly and cautiously. I need to buy some time until we can get the whole division back together and well supplied. I'm not sure what we are up against but I'm not going to waste the lives of my Marines on a reckless race to the Yalu River. I don't want my Marines home by Christmas in body bags." O.P. Smith was swimming against the tide and risking his career in the process.

Craig quickly replied, "I'll pass those orders on immediately and I'll breathe a lot easier when our regiments are back together."

"One other thing, Al. The railhead ends at Chinhung-ni, right?"

"Yes sir," replied Bowser.

"All right," said Smith. "Set up a large supply dump there. From there we can truck the supplies to Koto-ri, Hagaru, and Yudam-ni. We want to stock up on ammo, rations, medical supplies and fuel. We'll also need cold weather clothing, sleeping bags, tents and stoves. Have the supply people figure this out and then get it trucked up to these towns. Let's start building up these supply dumps. With winter coming, we may not be able to get resupplied by air and if something happens, we need to be ready. From what Litzenberg reports, winter may already be here. We need those supplies up there now!"

Craig left and Litzenberg and Murray soon got the word to slow it down. Bowser then began work on plans for resupply.

In the early afternoon, Smith visited his Marines at the division hospital. It was never an easy task but one that the men appreciated. They needed to know that their sacrifices and pain were valued. After his hospital visit, Smith went on to the newly established cemetery to pay his respects to those Marines who were killed. General Smith's uncharacteristic pessimism grew as he looked over the graves. He feared the number of casualties would only increase the further north they pushed and that this little cemetery might soon be vastly expanded.

"Dear God, protect my men," he prayed, "and guide me in this war."

Two days later a small landing strip was completed at Koto-ri that could only accommodate light observation planes. Smith thought that Hagaru, farther to the north, might be the best spot to build a landing strip long enough to handle the larger C-47 transport planes. This strip would be needed for resupplying and evacuating the dead and wounded.

Early on the morning of November 13, Smith flew by helicopter to Chinhung-ni, thirty-five miles to the north. The weather was bitterly cold and got worse the further north they went. At that time, helicopters could not fly up into the mountains because of problems with the gear boxes freezing up in the higher altitudes, so instead, Smith borrowed a jeep and along with two Marines drove the ten miles north to Koto-ri. He was stunned by what he saw. There was only a narrow one-lane road that led up into the mountains. It twisted and turned and could only accommodate one-way traffic. The hills were very steep on one side, sparsely covered with pine brush, scrub and trees, and on the other side of the road was a shear drop off, of anywhere from 400 to 1,000 feet. There were already patches of snow in places and ice on the road where springs overflowed it. The jeep slipped and slid over the road, making the short ride one that Smith would not soon forget. The further north he drove, the rougher, narrower and steeper the road became. As bad as this road had looked on maps and surveillance photos, it was much worse in person.

Smith wondered if tanks could make it up the road and how he was going to resupply his men over this torturous mountain terrain if it became impassable. He knew that the Chinese could easily block and cut off any traffic coming along this road. Between Chinhung-ni and Koto-ri, the road was barely wide enough to accommodate a truck. A few miles south of Koto-ri was Funchilin Pass where a one-lane bridge crossed over four giant pipes called penstocks. Water flowed downward from Chosin Reservoir through these pipes. If the Chinese blew the bridge, it would prevent any movement of wheeled or tracked vehicles. The road then wound its way down to the village of Koto-ri. It took Smith an hour to make the short trip from Chinhung-ni to Koto-ri and he thought to himself that Almond's planned operation was beyond risky; it was just plain foolish. He only wished he had General Almond at his side on this ride from hell. Maybe then he would see the light.

Smith was relieved when he finally arrived at Litzenberg's command post at Koto-ri. Smith and Litzenberg greeted each other warmly and Smith said, "That's a nasty piece of road. Litz, how are you doing?"

"I am just fine, sir, now that the Chinese have disappeared," replied Litzenberg.

"The reason for my visit Colonel, is that we've got new orders," started Smith. "Your regiment is to seize Hagaru. Litz, I want to reiterate what General Craig told you two days ago. I want you to proceed slowly and cautiously. You need to use your judgement regarding sound deployment of troops. I'm trying to buy some time so I can pull my regiments together. Also, Murray's regiment will follow you and protect the road from Chinhung-ni to Koto-ri."

"Yes, sir," answered the 7th Marine commander. "General, if something happens up here, we'll need more firepower. When can we get some tanks?"

"Litz, I'm concerned that the road might not be wide enough to handle our tanks but when I get back to Hungnam I'll talk to our engineer, Lieutenant Colonel Partridge. He is a wizard at these things and he'll get this squared away as quickly as possible."

Smith left and returned to his command post by jeep and helicopter, arriving before noon. As soon as he got there, he called in Lieutenant Colonel John Partridge, who was the commanding officer of the 1st Engineer Battalion. Partridge had graduated from the Naval Academy in

1944 and had years of engineering experience. During World War II he had served with the 4th Marine Division at Roi-Namur, Saipan, Tinian and Iwo Jima. He and his engineers would work minor miracles in the weeks ahead.

"John, I'd like you to make a recon of the road between Chinhung-ni and Koto-ri and give me a recommendation on what we have to do to improve it. It's pretty bad and I need to know if it's wide enough to handle our tanks and heavy equipment and if it isn't then how are you going to make it usable."

The two men conferred over the maps spread out before them and Smith recounted some of the problems he had encountered on the frozen drive he had taken earlier that day. He pulled notes from his pockets to show Partridge spots he had identified as potential problems.

Partridge left, got a jeep and a driver and headed to Chinhung-ni. From there he started up the one lane road into the mountains. It soon became apparent to Partridge that the hairpin turns were much too narrow for tanks. He knew that widening the road would be no easy task because of the steep mountain slope on one side and the sudden drop off on the other. There just wasn't much room to work with. When he arrived at Koto-ri, ten miles away he was stiff from the cold. After making some notes and sketches, Partridge then began the drive back to Hungnam.

It was late in the day when he reported back to General Smith. Smith invited Partridge to have dinner with him and General Craig in his small mess. Over a pleasant meal, the engineering officer began to thaw out and gave the generals a full report on improving that section of the road. "Sir, here is what we have to do. As you know, the road is very narrow and poorly drained. We'll start at the north section of the road by Koto-ri and work our way south. We'll have to do a lot of blasting and drilling and then the dozers and graders will get to work. Because of the melting snow we will have to put in culverts to keep the water off the road. If that road freezes, it would be hell to drive on."

"How long is this going to take John?" asked Smith.

"Sir, I think we can get most of this work completed by the eighteenth; at least we'll have enough of it done so tanks can move up the road."

"That sounds good, John," said Smith. "We need to get those tanks up there to protect the men."

The generals and the colonel finished their meal and then Partridge left.

Later that night, General Smith and General Craig talked.

"Eddie, I'd like you to go up to Hagaru tomorrow and have a look around. I think this might be a suitable location for a forward operating base for the division. Also, we need to figure out if we can build a landing strip up there."

"General," replied Craig, "I'll leave first thing in the morning."

On November 15, General Craig drove to Hagaru at the foot of Chosin Reservoir. Hagaru was a medium size Korean town that had been nearly flattened by bombing. Dominating the landscape was a large hill on the east side of the town. The unpaved streets were covered in snow. On the way into the town, Craig saw about 200 dead bodies laid in a row. The townspeople had been executed by the North Korean Communists and left behind as a barbaric warning to others: don't help the Americans. Craig was incensed by this ruthless slaughter of innocent civilians.

Like Smith and Partridge, Craig found the cold to be penetrating, despite winter gear. The wind seemed to roar down from Manchuria, making even a short time in the elements painful.

When General Craig returned from Hagaru later in the day, he reported to Smith.

"General Smith," said Craig, "you didn't exaggerate about either the state of the road or the cold. We have some real obstacles to overcome. I think that Hagaru would certainly be the best spot to have a forward base. Hagaru sits in a wide valley where we could build a landing strip and we could also store a lot of supplies up there. The town sits at a junction of roads leading to the north, northwest, and northeast. There are also a number of existing buildings at Hagaru still standing which could be used for medical facilities and other uses. Also two miles north of Hagaru is the hamlet of Sasu-ri. There's a sawmill there and it's full of fresh cut lumber. This will come in handy. Sasu-ri doesn't have much to offer; it's just a point on the map."

"Eddie, thanks for your input. I am going up there tomorrow to have a look and talk to Litzenberg. I can't shake the unease I am feeling about this operation. But at least we know we are doing our very best to keep the troops supplied."

Earlier that day, Smith had a frank conversation with Rear Admiral Morehouse, who was Chief of Staff to Admiral Joy. Morehouse had come to Smith's command post at Hungnam to get a first-hand account of what was going on. Smith found a sympathetic ear in this Navy officer. They agreed that there was a lack of realism in the Army plans and a tendency to ignore the enemy's capabilities.

"We can't forget that these Chinese soldiers are veterans of their internal wars," said Smith. "They are seasoned soldiers."

The more Smith thought about it, the more concerned he became. He hadn't been sleeping well the last few nights. Telling Litzenberg and Murray to slow down had been in direct contradiction to Almond's orders. Smith was a Marine who obeyed orders. He knew that was his job and what his career was based on. Smith knew dragging his feet with troop movement could put his career in jeopardy. There had to be a line where you could honorably follow your own judgement. But Smith wasn't sure exactly where this line was. He wanted someone to understand his concerns. But how? He tried talking to Lem Shepherd, in his regular chain of command, and he certainly had let Almond know what he thought of this folly. Neither man had been responsive. Either could relieve him tomorrow for disobeying orders. This kept Smith up at night.

This impending sense of doom that filled him caused Smith to again do something totally contrary to his military training. He went outside the chain of command and wrote a lengthy, detailed letter to the Commandant of the Marine Corps expressing his unease about the attack north. He had a good relationship with General Cates, the Commandant, and felt that he could share his concerns. Cates knew he wasn't a nervous Nelly and would not violate the chain of command without real reasons. He had met with Cates just after the Seoul battle, a few weeks earlier, and they had discussed the role the Army and its leadership were playing. He knew Cates was not totally enamored with MacArthur. After an hour of internal debate he wrote:

Although the Chinese have withdrawn to the north, I have not pressed Litzenberg to make any rapid advance to the Manchurian border. However, we are the left flank division of the Corps and our left flank is wide open. There is no unit of the 8th Army nearer than eighty miles to the southwest of Litzenberg.

I do not like the prospect of stringing out a Marine division along a single mountain road for 120 air miles from Hamhung to the border. I now have two regiments on this road and when Puller is relieved by the 3rd Infantry Division, I will close up behind him.

Smith went on to talk about resupply:

What concerns me considerably is my ability to supply two regiments in the mountains in winter weather. Air drop in winter is not a feasible means of supplying two regiments. Moreover, it will not provide for evacuation. The answer, of course, is to build a strip for C-119s and C-47s. At the altitude in which we are operating, the aviators require a 5,000 foot strip. X Corps thought it would be a fine idea if we built such a strip. With its other commitments, this is hardly a job within the capabilities of our Engineer Battalion. If we can find enough flat real estate in the vicinity of the reservoir to build a 5,000 foot strip, I will ask X Corps to give us a hand.

Even though the men who are up front are young and are equipped with parkas, shoe pacs and mountain sleeping bags, they are taking a beating.

Smith then talked about his lack of confidence in X Corps:

As I indicated when you were here, I have little confidence in the tactical judgment of X Corps or in the realism of their planning. My confidence has not yet been restored.

Time and again I have tried to tell the X Corps Commander that in a Marine division he has a powerful instrument but that it cannot help but lose its full effectiveness when dispersed.

Someone in high authority will have to make up his mind as to what is our goal. My mission is still to advance to the border. The 8th Army eighty miles to the southwest will not attack until the twentieth. I suppose their goal is the border. Manifestly we should not push on without regard to the 8th Army. We would simply get further out on a limb. If the 8th Army's push does not go, then the decision will have to be made as to what to do next. I believe a winter campaign in the mountains of North Korea is too much to ask of the American soldiers or Marines, and I doubt the feasibility of supplying troops in this area during the winter or providing for evacuation of sick and wounded.

Smith knew there were real risks in sending this letter. He trusted Cates to be discreet, but knew that once the letter was out of his hands, it was out of his control. His frank, unflattering assessment of X Corps' leadership could well come back to haunt him. Maybe it would happen right away here in Korea, or perhaps later in his career. But he strongly felt that he needed to be on record with his objections to the way this campaign was being waged. He finished the letter, mailed it and slept soundly that night.

It was an amazingly prophetic letter because General Smith seemed to know what was coming. He had a unique philosophy: be prepared for the worst and be optimistic when it comes.

At 8:00 in the morning the following day, Smith walked outside his command post. Waiting outside was a brand new Dodge station wagon painted Marine Corps green. Standing in front of the car was a sergeant with a Thompson submachine gun and a corporal holding an M1 rifle.

The men saluted and said, "Good morning, General."

Smith returned the salute. "Good morning."

Smith climbed in the front seat. The sergeant drove and the corporal sat behind Smith.

The sergeant asked, "Sir, we going to Hagaru?"

"Yes we are. Sergeant, where did you get this brand new station wagon? It's pretty fancy transportation."

"Sir, it sure is," the sergeant grinned. "The division got two of them yesterday and I thought since we're going up into the mountains that you'd be a lot warmer in this vehicle than a jeep."

"Thanks Sergeant. I did ride in a jeep earlier this week and I still haven't thawed out."

As they drove along the dirt and gravel road it was two lanes wide but when they reached Chinhung-ni forty-three miles later, the road narrowed to a single lane that wound its way up the mountains. Part way up, at one of the wide spots in the road, they came across an open jeep ahead of them and when they started to pass, Smith said, "Pull over Sergeant, I think that is General Harris."

And indeed it was Major General Field Harris, Commanding General of the 1st Marine Air Wing who was on his way to visit his son Lieutenant Colonel Bill Harris, who had become commanding officer of the 3rd Battalion 7th Marines, just six days earlier. Bill Harris would be serving under Colonel Litzenberg at Hagaru.

Not knowing the weather and road conditions, General Harris had taken an open jeep for the ride and he was obviously suffering from his decision. Smith opened the door of his heated vehicle and said, "Hey Field, would you like a ride?"

Field looked over and jumped out of his Jeep and said, "Hell yes."

Smith climbed out of the car and said, "Corporal, you hop up front, General Harris and I will ride in the back."

"God damn, this is more like it," said the nearly frozen general. "You wouldn't by chance have another one of these station wagons would you, O.P.?"

"Yes, in fact we do have one more."

With a grin on his face, Harris asked, "Can I have it?"

"You sure can."

"Thanks O.P. I really appreciate it," said Harris.

Ensconced in the comfortable vehicle, the two old friends caught up on news from Korea and home.

"O.P. does this operation make any sense to you?" asked Harris as the station wagon climbed the snaking road.

"Not a bit, Field. Putting a Marine division up in the mountains in the middle of winter just seems idiotic. I'm doing what I can to improve the situation as best I can. After your visit with your son, I want you to help me to find a spot for an airfield. It's the only way we can evacuate causalities and bring in fresh troops and supplies. It's bad enough that we are putting troops up here. We can't let them run out of food, fuel and ammo."

"I just want to say hello to Bill and congratulate him and then we'll find a location for our airstrip."

"Field, how is your son doing?"

"He's doing fine O.P. Thanks for asking."

Harris looked at Smith and said, "As you know, he was treated brutally as a POW in World War II."

"Where was he captured?" Smith asked.

"Bill was captured the first time when Corregidor was surrendered in May of 1942. He escaped with another American, Ed Whitcomb and they swam across Manila Bay at night. It took them eight and a half hours. O.P., can you imagine spending all that time in the water in the dark and what about the sharks? It makes me cringe. Well, somehow they made it ashore on the Bataan Peninsula."

Smith was listening intently. "Where did they go from there?" he asked.

"They headed north towards China through the jungle on foot, over mountains and sometimes by boat. They hooked up with some Filipino guerrillas and stayed with them for a few months. But when Bill heard that the Marines had landed on Guadalcanal, he stole a boat and headed towards Australia. O.P., he made it as far as some little island in Indonesia called Morotai. He was captured the second time when some civilians turned him in to the Japs. When the Japanese found out he was my son they sent him to a hell hole called Ofuna. It was a POW camp for high value prisoners. It was located outside of Yokohama, Japan."

Smith replied, "Field, I saw your son's service records and I know they treated him barbarically."

"O.P., Bill was a sharp kid, had a photographic memory and a fluent understanding of the Japanese language. He would sneak into the guards' office and look for information, including maps from Japanese newspapers. Back in his cell he would recreate this information by memory on scraps of paper and then hide it for future use. Well, one day a guard surprised him in his cell and found a small map in his hand. After a thorough search of his cell, the guard found Bill's stash of hand drawn maps, stolen newspaper clippings and a handmade dictionary of Japanese military terms. Then Bill was savagely beaten with a crutch taken from a fellow prisoner. He was beaten for forty-five minutes, long after he had lost consciousness. It took Bill a long, long time to recover."

"Field, it was terrible how they treated our men."

"I was happy when I learned that they had captured and tried those prison guards. The one that beat Bill so sadistically was caught after the war and sentenced to be hung. But you know, O.P. that is too easy a death for that bastard. I know that there has to be a special place in hell for that son of a bitch," said Harris.

"O.P., I don't know if you knew this but at the end of the war, Bill was given the honor of witnessing the Japanese surrender aboard the USS Missouri. He was one of four POWs," said Harris with obvious pride.

"Field, I'm proud to have Bill as part of my outfit," responded Smith.

"Thanks, O.P.," said Harris. "I've told him he was lucky to be under your command. By the way, nice job on the *TIME* magazine cover. It made us all happy to see your mug there and not Almond."

As they crossed the one lane bridge at Funchillin Pass, General Smith said to his driver, "Sergeant, pull over to the side of the road."

The driver eased the car off onto a narrow shoulder from where you could see the entire valley below.

Smith looked over at General Harris and said, "Let's get out of the car for a second. I want to show you something."

They left the car and were then greeted by a blast of cold air. Both men shuddered.

General Smith looked at the bridge and then over to General Harris. "Field, why is this bridge still here? We're attacking north and fighting the Chinese or the North Koreans or the Chinese volunteers or whatever Almond wants to call them."

Smith added with a little sarcasm in his voice, "They are retreating north across this bridge and wouldn't you think that the first thing that they would do is blow this bridge to keep us from pursuing them?"

"I agree O.P. It doesn't make any sense at all."

"I wish I could answer that question but I can't and that bothers me," added Smith.

The generals quickly climbed back into the warm station wagon. The car continued forward over the pass then down into Koto-ri.

As they drove through the little town, General Harris looked at Smith and commented, "O.P., this is a bleak looking place."

General Smith pointed out, "Field, you can see the 2nd Battalion of Murray's 5th Marines have already set up camp there. It sure looks like the only flat ground anywhere near here. We were lucky to find space for the small airfield, but we can't use Koto-ri for the big airstrip."

As the ride continued Smith and Harris talked about the feasibility of building an airfield in this cold weather in Hagaru.

"It's not just the construction part, O.P. Flying at this high altitude adds a new layer of risk for the pilots and crew as well as for any passengers or wounded that they are evacuating," said Harris. "The elevation there is 3,500 feet and that requires a longer runway because of the decreased density in the air. This reduces lift and engine power."

"Okay, I can understand that," replied Smith. "That's a lot different than when we fought in the Pacific where we were always at sea level."

When they got to Hagaru, General Smith's driver asked for directions to the headquarters tent of the 3rd Battalion, 7th Marines. Later, they pulled up to the command tent. The generals got out of the warm car and were hit by a gust of Arctic air. They quickly entered the command post. The tent was chilly but not like the frigid air outside. Bill Harris was sitting behind a field desk wearing a coat and gloves. When he looked up, he was pleasantly surprised to see not only his father, but also General Smith.

Harris quickly stood and said, "Dad, what are you doing here? And General Smith, good morning sir."

General Harris walked up to his son, warmly shook his hand and said, "Bill, congratulations on your promotion."

"Thanks Dad. It is great to see you."

O.P. Smith stepped forward and said, "Bill, let me also congratulate you," shaking his hand firmly.

Then Bill Harris asked, "What are you two doing up here? Surely you're not here just to see me."

Field Harris said, "Actually, I am. And then General Smith and I are going to scout out a spot for an airfield."

General Smith said, "I'll let you two catch up. Field, when you're through, I'll meet you at Litzenberg's command post. Bill, it was great seeing you and would you please have someone guide me to Colonel Litzenberg?"

Bill Harris called to one of his sergeants, "Sergeant would you please show General Smith to Colonel Litzenberg's headquarters?"

"Thanks Bill," replied Smith as the general and the sergeant walked out into the bitter cold.

Twenty minutes later, General Harris met up with General Smith and Litzenberg. Litzenberg smiled, extended his hand and said, "General Harris, it's a pleasure to meet you sir. And welcome to Hagaru. I am really glad that Bill took over my 3rd Battalion."

Harris replied, "It was a great visit and I'm so happy that I got a chance to see Bill and congratulate him on his promotion. I just wish his mother and wife were able to see him now."

After a brief update in the command tent, Smith and Harris left Litzenberg, hopped back into the Dodge station wagon and began looking around the valley of Hagaru that Litzenberg had described, in search of an appropriate spot for a landing strip. Just south of town they came across a bean field which appeared to be frozen. Litzenberg had thought this might be a good spot and already had his men test it to confirm that the ground was frozen to a depth of 16 inches. Harris felt this surface would be strong enough to land his planes. It wasn't perfect but with a lot of work they might be able to stretch it to the 4,000 feet that Harris needed to land his large C-47 planes.

"Now the big question," Harris pondered. "Do you have the resources to be able to smooth this thing out and make it long enough to handle two-engine planes?"

Smith replied, "I'm going to ask Almond if X Corps can build this strip. It seems to me that this should be their responsibility but as you have probably heard, General Almond and I are not always in agreement."

As they left the bean field and started the drive back, Harris said to General Smith, "O.P., you and I have been through a lot in World War II and if this Korean thing gets bigger, you're sure going to have your hands full. I want you to know that all my resources are at your disposal. I will muster every plane I have to support your men on the ground. But believe me, the weather and terrain are going to factor in on what air power we can provide."

Smith nodded and said, "I think we got a lot accomplished today and I'm feeling a little bit better about taking care of some of my big concerns. This airfield will mean everything should we get into a full-scale battle. We just have to get Almond moving on this construction and then pray for good weather."

"O.P. what is the scuttlebutt? What is your take on how big this thing is going to get? From the looks of it and what you read in the press, we have the North Koreans on the run, but you have to be worrying about whether or not the Chinese are going to get involved. I know that MacArthur does not believe the Chinese will get into the fight and that Almond is really pushing hard to get all of X Corps up to the Chinese border. I'm afraid that all hell could break loose and you would be stuck up here in these mountains. These North Korean winters are vicious as we already know. Christ, I damn near froze to death in that Jeep before you rescued me in this station wagon."

"Between you and me, Field, I have been trying to stall and slow down our advance. I want to get all of my Marines together so we are a powerful fighting force. I talked about this with Shepherd and he just told me to try and stay with the program, and that Almond's a good guy, just doing his job. Field, I think this is lunacy; spreading out our troops so far between each other and not having the necessary supply dumps and airfields to carry out an operation. I will certainly do my job but I am going to take every precaution not to waste the lives of my Marines. If they relieve me, it won't be because I acted rashly!"

The two generals continued their ride back home and reminisced about old times. They had known each other since 1946 when they had been appointed by the Commandant to serve on a board to review the concept of amphibious warfare in the atomic age. "I wish all those nay-sayers who were sure we would never again have any need for amphibious landings had been around at Inchon," said Smith.

"Your men sure showed them how it's done," replied Harris.

"It was great to have some time to visit with you, Field," said Smith as they neared their destination. "I hope I'll have a chance to see you again soon. I know that all of the troops have been told that we will be home by Christmas, but I have a bad feeling about this. This thing could get a lot bigger and I don't like this terrain and I don't like the weather. All of the veteran troops that we have were in tropical conditions in World War II and I wonder if we are prepared both mentally and physically to fight in this type of extreme cold. Remember how hot we all were in World War II? How the men swore that if they ever got home, they were moving to Maine or Minnesota? Well yesterday I heard a Marine saying that he was so cold in Korea that when he got home he was moving to Arizona or Florida. He said he had relatives in Buffalo that he would never visit again, even in the summer! Cold will numb your spirit as well as your body. And spirit is what makes a man a Marine."

Smith dropped General Harris off at Chinhung-ni where his helicopter was waiting for him. As good-byes were exchanged, Smith smiled as he told Harris that his new, heated station wagon would be delivered to him first thing in the morning.

It was well after dark when Smith's car pulled up to his CP at Hungnam. As he walked through the door, he was surprised to once again see Major General Lowe. Lowe shook Smith's hand and said, "Good to see you again O.P. I hear you took a trip up to Hagaru today. How did that go?"

Smith smiled, happy to see Lowe again. "Frank, have you eaten yet? How about joining Eddie Craig and me for some dinner. I'm famished. We can talk about Hagaru as we eat."

The three generals sat and were served their dinner. General Smith spoke first.

"Frank, on the way up to Hagaru, I ran into Field Harris and we rode together the rest of the way. We found and agreed on a spot to build an airstrip. Litzenberg had scoped out the spot earlier and it looks like it will work. I'm going to see if X Corps can take on this job. My engineers are busy working on the road north. I really feel better now that we have those plans in place. And how about you Frank? What brings you back here?"

"O.P., I just flew in from western Korea where I was visiting with the Army's 1st Calvary Division. I brought some map overlays with me that show the dispositions of the 8th Army. I'm going to turn these over to the people at X Corps. I thought that you should see them as well."

Eddie Craig said, "Frank, we don't have any idea about what's happening over there or even where the Army is. There is just no communication between us."

"I know just what you mean Eddie. They tell me the same thing. It doesn't make a damn bit of sense to me. Hell, I'm the only physical liaison between the 8th Army and X Corps. All communications have to go through Tokyo. The Army on the west side doesn't know what's happening over here or where the Marines are," said an exasperated General Lowe.

O.P. Smith put down his fork and said, "Frank, this really bothers me because our left flank is wide open."

"I would agree with you O.P., especially after what happened to the Army at Unsan," said General Lowe.

"As you know, gentlemen," continued Lowe, "I have spent a lot of time with your Marines. I wrote the President and I told him that the 1st Marine Division is the most efficient and courageous combat unit I have ever seen or heard of. And I also told him that the safest place in Korea was with a platoon of Marines. And I mean it!"

Both Marine generals smiled and O.P. Smith said, "Thanks Frank. That's a great compliment to all our Marines and we appreciate your sharing that sentiment with President Truman."

What Lowe didn't say however, was that he had also written the President about his impressions of Smith describing him as an "offensive tiger." He also went on to say that Smith's officers and men idolize him although he is a strict disciplinarian—Marine discipline!

On November 17, Smith told Litzenberg to move up to the west side of Chosin Reservoir to Yudam-ni and Murray to move to the east side of the reservoir to Sinhung-ni.

Smith later flew to meet his old friend Rear Admiral Jimmy Doyle on the USS Mount McKinley. Smith recalled how hot he had been when living on this ship in the summer. Those days were clearly gone. Over dinner the two friends shared their concerns about the coming action. It was a nice break for Smith.

The Marine engineers were truly miracle workers. By the eighteenth, as promised, they had improved the road so tanks could be sent forward to hook up with Litzenberg. Smith had asked Almond to provide him with X Corps engineers to build the airstrip at Hagaru but Almond told him he'd have to use his own engineers.

Smith had explained the need to have planes that could evacuate wounded and Almond had responded, "General, I think you are completely overstating the number of casualties here. We have the North Koreans on the run."

Smith fumed, but wasn't surprised. So now with the road open, it was up to Lieutenant Colonel John Partridge's 1st Engineer Battalion to construct the airstrip.

Five large Caterpillar bulldozers chugged up the newly widened road and arrived on the nineteenth to begin construction of the runway. As the temperatures continued to drop, the heavy equipment couldn't penetrate the frozen ground. Marine engineers welded steel teeth onto the dozer blades so they could tear into the ground. Partridge's engineering plans called for an airstrip of 3,200 feet. The problem was that the engineering field manual called for a runway of 4,000 feet because of the high altitude of Hagaru. Everyone hoped and prayed that the shorter runway would work. Construction of the airfield would proceed twenty-four hours a day with the engineers working in the frigid temperatures under flood lights at night.

Another big problem that Smith had to tackle was operating helicopters and light aircraft at high altitudes in cold weather. On the nineteenth, Smith drove to Yonpo airfield a few miles south of Hungnam. Here he was met by Major Vincent J. Gottschalk, the squadron commander of the Marine Observation Squadron, which was under Smith's operational control. Gottschalk had solved the problem of flying helicopters in the cold at high altitudes. First, he used lighter grease and oil. And when possible, helicopters would be placed in hangers at night. Next, he equipped all helicopters with metal blades. Prior to this, most helicopters were using fabric blades.

From helicopter issues, Smith's day moved to international diplomacy. The Korean War was a United Nations action, although the U.S. forces comprised the vast majority of combat troops.

Lieutenant Colonel Douglas Drysdale and his 41st Independent Commandos, British Royal Marines arrived at Hungnam on the morning of November 20. Earlier in November, Vice Admiral Turner Joy, Commander Naval Forces Far East, had asked Smith if he could make use of this outfit of Royal Marines. Joy indicated they were hoping to serve with the 1st Marine Division. They did not want to be attached to an Army unit. Smith appreciated this sentiment and replied that he would be glad to have these fine troops and that he would put them to good use. This British unit was made up of fourteen officers and 221 enlisted men.

At 9:40 am on the same day, General Smith flew by helicopter from Hungnam to visit the 7th Marines at Hagaru. Litzenberg reported to Smith that the Marine engineers were making good progress on the construction of the airfield. Colonel Murray, who was temporarily headquartered at Hagaru, while his 5th Marines were heading to the east side of Chosin Reservoir, stopped by the command tent to see Smith.

Both Litzenberg and Murray shared their concern about the reports of Chinese in their area. Civilians and POWs were supplying intelligence that the Chinese were massing in great numbers and that they often took shelter in caves when planes flew overhead. These reports from his regimental commanders only reinforced Smith's belief that the Chinese had actually entered the war. He had zero confidence in X Corps' intelligence or MacArthur's for that matter. Smith was worried that there might be an entire Chinese Army hiding in these mountains. He had over 15,000 men strung out over miles. How many Chinese could there be?

On the twenty-first, the Army's 3rd Infantry Division took over Puller's position around Huksu-ri and now he was able to move his regiment north behind Litzenberg and Murray. Smith breathed a sigh of relief when Craig gave him the news.

"I am feeling a little better about this whole operation. The Division is still strung out, but not so far apart now," said Smith. "We still have no real idea how many Chinese are out there, but at least the Marines are not so widely separated and our supply depots are taking shape. God willing, the airstrip will do the job and then maybe I could sleep a little sounder," said Smith.

"I agree, General," said Craig. "This is the stuff that causes ulcers."

Politics and glad handing were never part of the job that Smith enjoyed. When the Secretary of the Navy arrived on November 22, the general decided to take him to the division hospital to visit the wounded. That somber setting made small talk unnecessary.

Thursday, November 23, was Thanksgiving Day and Litzenberg's 7th Marines moved north to Yudam-ni while Murray's 5th Marines continued their move to the east side of the reservoir.

Smith spent the day in Hungnam. Although his friend, Admiral Doyle had sent a cooked turkey to his kitchen, Smith had been required to attend a lavish dinner hosted by Almond at the X Corps headquarters. Smith thought that the real linens, cocktail bar, china, silverware and place cards were over the top in a war zone. By contrast, his Marines in the field were eating their turkey dinner out of their mess kits. And given the weather conditions, the turkey was not likely to be hot.

At the dinner, he did have a chance to talk with Generals Barr and Field Harris. Barr's 17th Infantry had reached the Yalu River the day before. Smith had sent him a congratulatory dispatch, but at the dinner party, Barr told Smith that the whole operation had been half-assed.

"We operated on a shoe string," Barr said. "We only had a single day's supply of food and ammo on hand. I know you have been working on keeping your men well supplied. That's the smart play, O.P. You don't want them in this desolate land without supplies."

"I agree," said Smith.

General MacArthur had come to Korea on November 24 to see the start of the 8th Army's offensive which MacArthur asserted would "end the war." His press announcements promised that the war should be won in a couple of weeks and Almond got his orders to execute the planned attacks.

Puller's 1st Battalion arrived at Chinhung-ni on Thanksgiving Day. Two days later on the twenty-fifth, his regimental headquarters and his 2nd Battalion moved ten miles north and took over Koto-ri. Two companies of his 3rd Battalion arrived at Hagaru, eleven miles north of Koto-ri on November 26, under the command of Lieutenant Colonel Thomas Ridge. Puller's job was to keep the road open from Chinhung-ni to Hagaru. It was a twenty-one mile stretch of mountainous road and defending it would be no easy task.

Almond wanted the 7th Marines and the 5th Marines to attack north from Yudam-ni but to do that he had to replace the 5th Marines on the east side of the reservoir. The Army's 31st Infantry Regiment from the 7th Division, under the command of Colonel Allan MacLean was given this job. This unit was put together at the last minute and because of the lack of transportation and travel distances it would arrive at different times over the next few days.

On the twenty-fifth, Smith got word that Murray's 5th Marines were relieved by the Army's Lieutenant Colonel Faith and his battalion, the lead unit of the 31st Regiment. This left Murray free to move his regiment from the east side of Chosin Reservoir to the west side and hook up with Litzenberg's 7th Marines at Yudam-ni.

Faith set up his command post in a hut on the lower slope of Hill 1221. The military typically named hills according to their elevation in meters. Murray advised Faith not to move farther north without orders from the 7th Army Division. "Dig in," Murray advised. "Be careful and don't let your troops get too strung out. I don't know what is north of us and intelligence indicates there are plenty of Chinese soldiers around here." But Faith was pressed by Almond to attack north. The Yalu was seventy-five miles north and Faith was a leader with limited tactical skill or combat experience. His father had been a brigadier general so he was raised in the Army. During World War II, Faith served as an aide to General Matthew Ridgeway for more than three years. He had never commanded any troops in battle before the Inchon landing. When Murray departed, the only direct communication between Faith and the 1st Marine Division

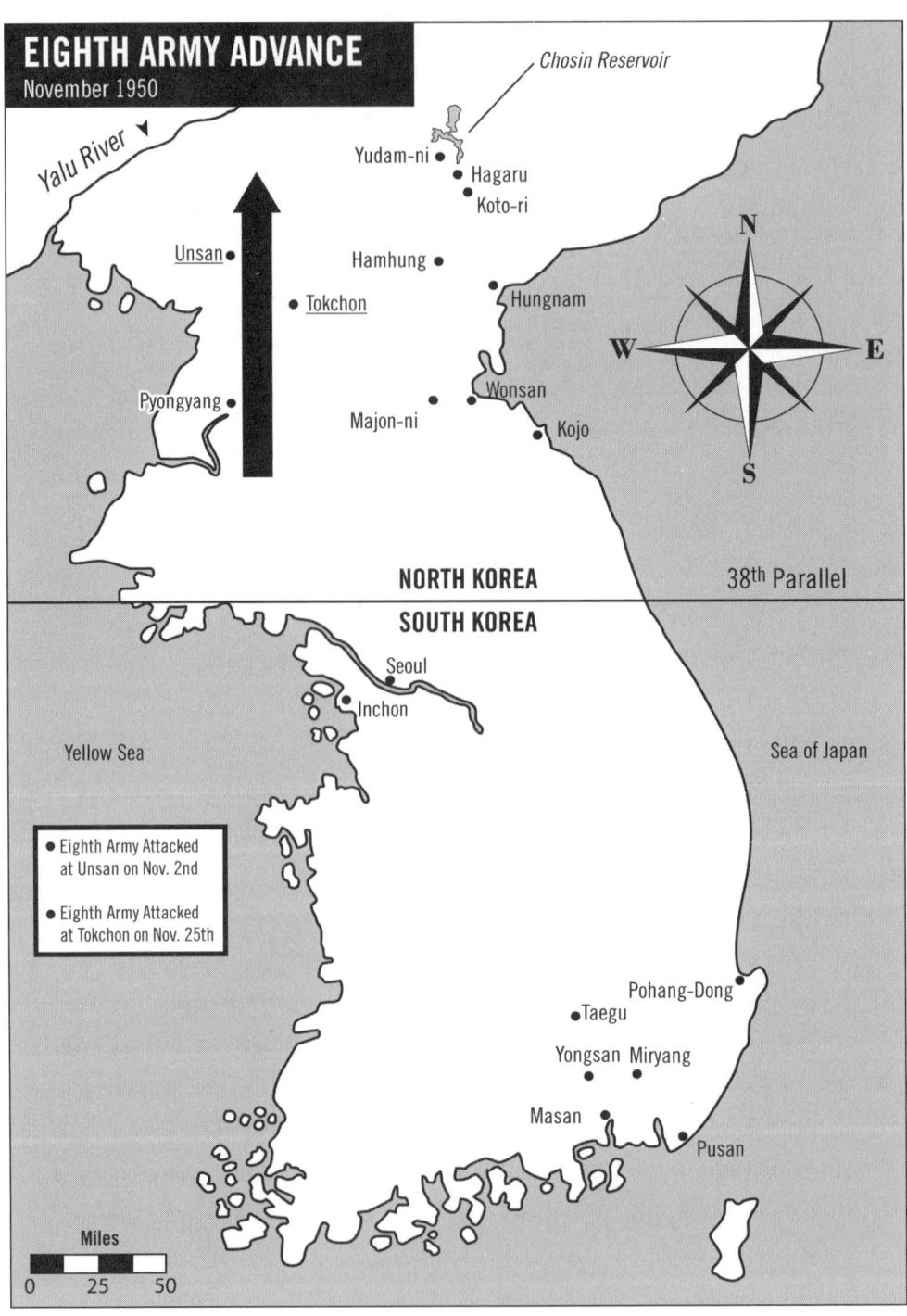

EIGHTH ARMY ADVANCE
November 1950

Chosin Reservoir

Yalu River

Yudam-ni
Hagaru
Koto-ri

Unsan

Hamhung

Tokchon

Hungnam

Pyongyang

Majon-ni

Wonsan

Kojo

N
W E
S

NORTH KOREA   38th Parallel
SOUTH KOREA

Seoul
Inchon

Yellow Sea   Sea of Japan

● Eighth Army Attacked
  at Unsan on Nov. 2nd

● Eighth Army Attacked
  at Tokchon on Nov. 25th

Pohang-Dong
Taegu
Yongsan Miryang
Masan
Pusan

Miles
0   25   50

would be provided by his attached tactical air control team led by Marine Captain Edward Stamford. That communication was accomplished by relaying information from the ground to an aircraft and then back to the 1st Marine Division. Stamford had been attached to Faith's battalion since the Inchon landing. His job was to call in aerial strikes during the battle. And in order to do this, he had to be on the front lines.

On Sunday morning, November 26, Smith received a message from X Corps that the 8th Army had come under heavy attack in the west at a town called Tokchon. The message went on to say that details would follow when available. Smith briefly discussed these latest developments with Craig and then he flew from Hungnam to Yudam-ni by helicopter to visit Litzenberg.

Smith flew up over the road looking for any kind of Chinese activity but saw none. His instincts however told him that something was wrong. The Chinese had disappeared after their battle with the Marines on November 2. Where were they? There were miles of wooded mountains out there, but could they really hide a Chinese army?

Smith landed at what he thought was Litzenberg's CP but it was actually the CP of Lieutenant Colonel Ray Davis, the CO of the 1st Battalion, 7th Marines. Davis and Smith talked for a while and then Davis told him how to get to Litzenberg's CP which was 5,000 yards south.

The general climbed back into his helicopter and the pilot took off. A minute later they spotted Litzenberg's CP. As they were coming in for a landing there was no air moving and when the helicopter got within ten feet of the ground it dropped like a rock. The passenger, pilot and helicopter took quite a jolt but were unharmed.

When Smith eased himself out of the helicopter, Litzenberg was there to meet him. He said to Smith, "General, that was a pretty grand entrance." Smith did not appreciate the humor and just shook his head. Helicopters were new to warfare and Smith had not completely adjusted to this mode of transportation.

Litzenberg's jeep drove them to his command post which was a large heated tent. It was zero degrees outside and the heated tent did little to warm the men.

Smith spoke first. "Litz, this whole operation smells bad. Flying up here I saw no enemy activity. In fact, all the air reconnaissance for the last few days has turned up absolutely nothing. They report no sign of the Chinese. Something just doesn't feel right. Litz, anything new to report?"

"Yes sir. We captured three Chinese soldiers this morning. They claim there are eight Chinese divisions hiding in these mountains. They told us their objective is to move south against us and cut the road. Their goal is to annihilate us. Specifically, their goal is to annihilate the 1st Marine Division."

"Do you believe them?" asked Smith.

"General, something is going on here. It's too quiet. You know that when you have been in combat situations like we all have, I think you develop a kind of sixth sense. It's unexplainable, but it's real. The civilians who live around here are also reporting a lot of Chinese up in the mountains; they are taking their food and forcing them to act as guides. The deer are moving down the ridges. Something is spooking them. Our patrols have also run into some enemy resistance, so there is no doubt they are out there. I think we have to pay attention to this intelligence."

"Litz, before I left this morning, I received some very limited information that the Chinese attacked the 8th Army on the west coast. I don't have any details yet, but I wanted to give you a heads up on this. Yesterday I had a meeting with General Almond at X Corps. Our orders have changed. Now, we are to advance west to hook up with the 8th Army. We will be the right arm of a pincer and the 8th Army will be the left. Our attack is supposed to start tomorrow on November 27 and Murray's regiment will lead it. You will follow. But until we know what the devil is happening in the west with this news of a Chinese attack, I don't want us as the only side of a non-existent pincer. "

Litzenberg looked grim. "I sure as hell hope we don't end up fighting the Chinese up here. This cold weather is bad enough but to do battle in it would be pure hell on the troops."

"Litz, do you have enough tents and stoves?"

"Yes sir. Each platoon has a tent and a stove. These men can put one up in about fifteen to twenty minutes. The men don't sleep in these tents but we rotate six men at a time through the warming tents. It gives them about twenty minutes to get warm and get something hot to drink."

Smith replied, "Marines have never fought in this type of weather but we have to be prepared for it, and we will be. One more thing Litz, the high ground between Hagaru and Yudam-ni is Toktong Pass. Put a company of Marines up there just in case something happens. I want to control that ground. It could be a very strategic location."

"Aye aye, sir."

Smith left and flew back to Hungnam. He could see for several miles on either side of the road but still there was no sign of the Chinese. Where were they, he wondered? Eight Chinese divisions would be about 80,000 men. How do you hide 80,000 armed men? Smith had driven this road twice and had flown over it twice. He probably knew as much about this road as any officer in his division. If the Chinese attacked, he knew that he would have a very difficult time keeping it open. This scared the daylights out of him.

When he arrived at the division CP, he was met by Colonel Bowser.

"General, we have more information about the 8th Army. It looks like it's a major disaster over there in the west. The 8th Army is starting to retreat. I don't know what they're up against but it looks like it's turning into a rout."

"Al, keep me informed of any developments over there."

A short time later, a corporal entered General Craig's office and handed him a message from the American Red Cross. Craig opened it and read the first line. 'Father not expected to live...'

He was so shocked, his eyes started to water. He immediately felt so conflicted. He had to see his father one more time, however because of the current tactical situation, he knew that General Smith needed him.

Craig took a moment then stood up and walked down the hall to General Smith's office. He entered and asked, "General, do you have a minute?"

"Sure Eddie, have a seat. What's up?"

"Sir, I just received a message from the Red Cross. My father's dying."

General Smith's thoughts immediately returned to the many evening's conversations that he and Eddie had. The two generals had now been together for two months in Korea and had become close.

Frequently after dinner, they would sit and review the events of the day. Often the conversation would turn to family. From these discussions, Smith knew that Eddie Craig was very close to his father, a well-known Army doctor. Smith was also aware that in World War II, Craig's commander had refused him emergency leave to visit his dying wife.

Smith was stunned. This news couldn't have come at a worse time. Could he risk sending General Craig home? He had counted on Craig to take over in the command post at Yudam-ni and coordinate Murray and Litzenberg's operations. Smith looked at Craig for what seemed to be a long time. He had a decision to make and it would not be an easy one. Eddie Craig was an outstanding Marine and his trusted assistant division commander. In view of what was happening to the Army in the west and all the uncertainty in Smith's area of operations, he needed Craig more than ever.

Finally he said, "Eddie, I'm going to send a message to General Shepherd in Hawaii that if I do not receive word to the contrary, I plan to grant you emergency leave. You need to get there. Start packing."

Eddie Craig stood up, looked at Smith and said, "Thank you sir. I know that this was a tough decision for you. I'm sorry to have added to your burdens."

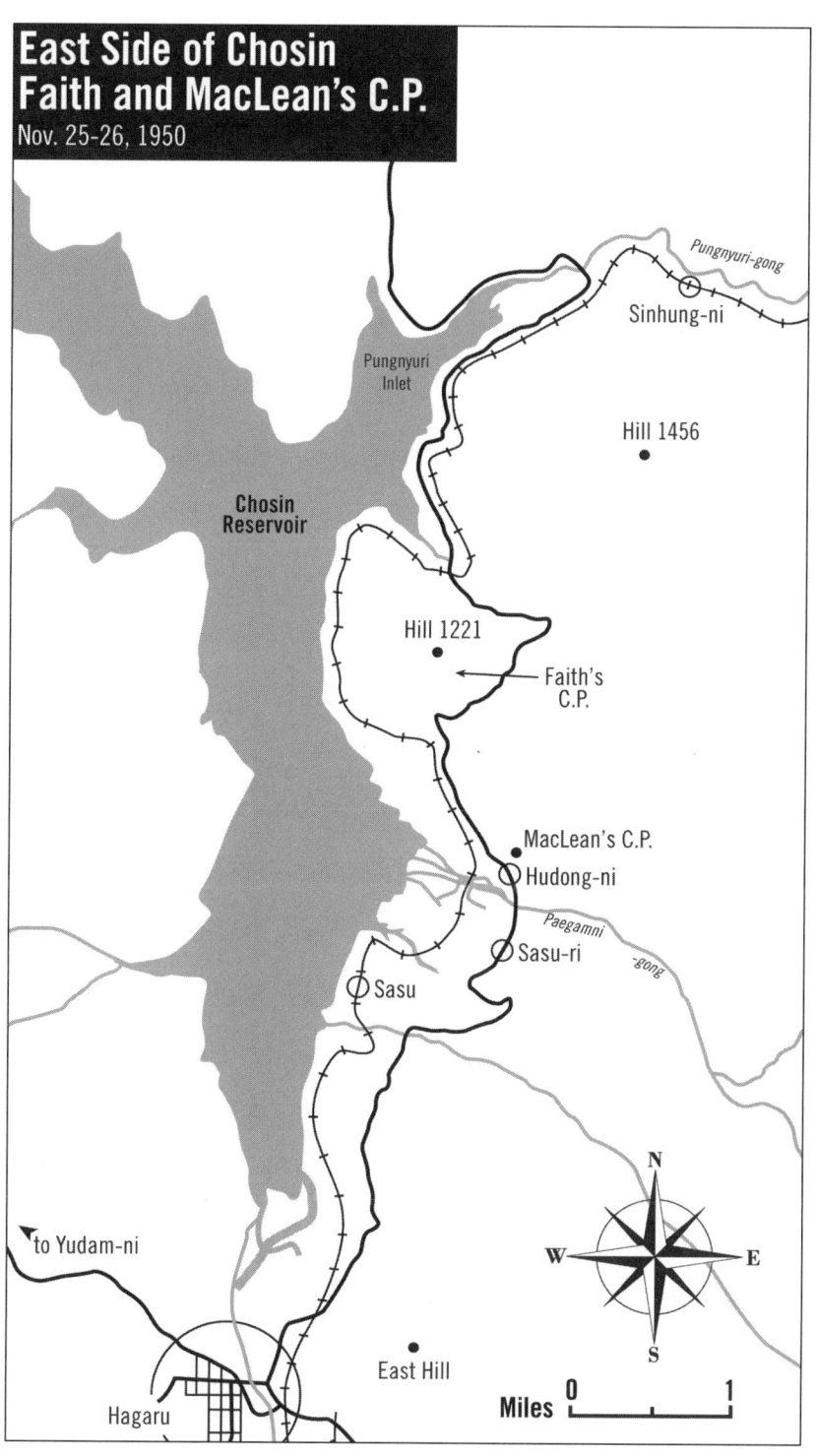

East Side of Chosin
Faith and MacLean's C.P.
Nov. 25-26, 1950

Pungnyuri-gong

Sinhung-ni

Pungnyuri Inlet

Hill 1456

Chosin Reservoir

Hill 1221

Faith's C.P.

MacLean's C.P.

Hudong-ni

Paegamni -gong

Sasu-ri

Sasu

to Yudam-ni

N
W   E
S

East Hill

Hagaru

Miles 0   1

When Craig left, General Smith called Bowser into his office and informed him that General Craig would be going on emergency leave because his father was dying.

Bowser said, "I know General Craig is very close to his father. It's the right thing to do."

"I'm glad you think so," said Smith, "because I imagine I'll take some heat on this one."

EAST SIDE OF CHOSIN

Shortly before noon on the twenty-sixth, on the east side of the reservoir, Army Brigadier General Henry Hodes, the assistant division commander of the 7th Division, arrived at Lieutenant Colonel Faith's command post. Hodes was known as "Hammering Hank." He had fought bravely at Omaha Beach and the Siegfried line in Europe. Hodes had been assigned by General Barr, 7th Division Commander, to direct the newly formed 31st Regiment's operations.

Faith was finally informed of his responsibilities by General Hodes. Lieutenant Colonel Faith's battalion was to be joined shortly by the balance of Colonel MacLean's 31st Regiment. This newly formed 31st Regiment of the 7th Division would occupy the east side of Chosin Reservoir until further orders to attack north to the Yalu River.

By mid-morning, Colonel Allan MacLean arrived, much to Faith's relief, to take command. Like Faith, MacLean's first real combat leadership experience had been weeks earlier at Seoul. He was forty-three years old and a big, burly man who had played tackle on the West Point football team. Colonel MacLean informed Faith that the 3rd Battalion would arrive the next day and that his newly formed 31st Regiment would attack north as soon as they were assembled. Faith started to move his battalion north of the Pungnyuri Inlet with MacLean's approval even though Colonel Murray and his departing Marines had just the day before strongly cautioned against being too strung out un-

The Savior

til all the Army's troops had arrived. Faith and MacLean both knew that General Almond wanted to be able to report to MacArthur that the Army was at the Yalu.

MacLean's regimental command post was set up in a school house in Hudong-ni, a village one mile south of Hill 1221.

For the Marine and Army units around the reservoir, this would be their last quiet night. This was the calm before the storm.

7

THE CHINESE

Mao Tse-tung, Chairman of the Chinese Communist Revolutionary Party, had been surprised by the Inchon landing. In a November 12, 1950, communique with General Peng, Commander of all Chinese Communist Armies, Mao stated: "It is said that the American Marines' 1st Division has the highest combat effectiveness in the American Armed Forces. General Sung should make the destruction of the 1st Marine Division, as the strongest of American Divisions, his main effort." Mao's plan was not to just push the UN forces back to South Korea. He wanted them destroyed.

As the 1st Marine Division was moving north, General Sung Shin-lun, who commanded the Ninth Army Group, was leading his twelve divisions south to fight the Marines. He would hold four of those divisions in reserve. He had the responsibility for the area of northeast Korea. Sung was only forty years old but had been leading men in combat since he was seventeen. Like Chairman Mao, Sung was from the Hunan Province. He graduated from the prestigious Whanpoo Military Academy. He commanded a regiment on the Long March, was a master at guerrilla warfare and was known for his bravery in battle. He was considered one of Mao's finest generals and his orders from above were crystal clear: annihilate the 1st Marine Division.

Sung's strategy was first to delay the Marines from moving north which would give him time to amass his forces. He had already successfully employed this tactic at Sudong on November 2, when he attacked Litzenberg's 7th Marine Regiment, inflicting heavy casualties. The second part of Sung's strategy was then to let the Marines advance deep into North Korea, isolate them and then crush them. Almond seemed to be playing right into Sung's hands.

A Chinese division was made up of 10,000 men; each division had three infantry regiments comprised of about 3,300 soldiers. At the company level, the troops would generally all be from the same town and ethnic group. The key advantage Sung possessed was numbers. He was commanding about 120,000 Chinese soldiers. Once a Chinese peasant became a soldier, he was a soldier for life. There were no provisions for honorable discharge. He stayed there until he was killed, captured, deserted or was so badly wounded that he could never fight again. He was told that U.S. forces were marching toward the Yalu River and planned to invade Manchuria. He was also told that any Chinese soldier surrendering would be immediately decapitated by the Americans.

The Chinese soldier was used to suffering great hardship and had the ability to march and fight with very little food. He typically had just enough to last for five days. Each soldier carried these rations in a cloth roll slung over his shoulders. There was no variety in his diet. He either ate rice from home or a ground up powder consisting of rice, millet and peas. If he couldn't heat it up, he would merely mix it with water and eat it cold. Their main source of resupply was to take what they could from the houses of local North Korean peasants. That meant the peasants would starve, but such were the fortunes of war.

Most of General Sung's troops had been stationed in southern China where temperatures were moderate. They had been sent north by rail and then marched 150 miles, beginning in mid-October to North Korea.

The Chinese soldier was also poorly clothed. His uniform was suited for the southern China posts but not for the Siberian cold of

North Korea. He wore a heavy quilted cotton uniform over his summer uniform: one side was mustard yellow and the other side was white for winter wear. He had a heavy cotton cap with fur lined ear flaps. His shoes were canvas with crepe rubber soles. When these "tennis shoes" wore out, the men would wrap their feet in rags. The Chinese soldier rarely had gloves or mittens and had to resort to tucking his hands into the sleeves of his jacket for warmth.

The Chinese army had no air or heavy artillery support. The Russians had reneged on an agreement with the Chinese to provide air support for this war. Stalin in Russia had decided that a strong China was not in his best interest. The Russians did supply some arms, but not decisive support. The Chinese also lacked tanks and logistical support. They often relied on coolies to transport supplies. Their medical support was nearly non-existent. A key part of their battle plans was to overcome the enemy and take all his weapons and supplies. The ground soldier had a mixture of weapons of Russian, British, Japanese and American origin. The United States had sold millions of dollars of rifles, machine guns, bazookas and other arms to the Chinese Nationalists after World War II at fire sale prices. Unfortunately, after the Chinese Nationalists were defeated by the Communists in 1949, most of these weapons ended up in Communists hands. They were also armed with small mortars and hand grenades. The cold affected the Chinese weapons just as it impacted those of the United States.

Sung told his men, "Soon we will meet the American Marines in battle. We will destroy them. When they are defeated the enemy army will collapse and our country will be free from the threat of aggression. Kill those Marines as you would snakes in your home."

Sung's headquarters was well hidden in the mountains ten miles north of Yudam-ni. From here he would direct operations against the Marines in the Chosin Reservoir area.

8

THE NEWS IN AMERICA

During November of 1950, there were two events in the United States that shifted Americans' attentions from the Korean War: the failed assassination attempt of President Truman and the "storm of the century."

On November 1, President Truman was living in the Blair House while parts of the White House were being remodeled. Shortly after 2:00 pm, two Puerto Rican Nationalists, Oscar Collazo and Griselio Torresola, angry at the United States for its treatment of their homeland, tried to assassinate the president.

Torresola approached the Blair House from the west while Collazo walked up behind Capitol police officer, Donald Birdzell who was standing on the steps of the Blair House. While President Truman was taking a nap on the second floor, Collazo tried to shoot Birdzell, but he forgot to chamber a round and the gun did not fire. Collazo quickly got the gun to work and fired his weapon just as Birdzell was turning to face him and shot the policeman in the right knee. Reacting quickly, Birdzell limped out into the street to draw fire away from the Blair House. While he moved, Collazo kept shooting and hit him again.

Hearing gunshots, Secret Service Agent Vincent Mroz ran out of the east side of the Blair House, opened fire, and shot Collazo in the chest.

At the same time, Torresola on the west side of the house caught White House police officer Leslie Coffelt by surprise shooting him four times. Torresola then shot another police officer, Joseph Downs in the hip and then in the back and neck. Downs somehow got in to the basement and locked the door preventing Torresola from entering the Blair House. Torresola then turned and shot Birdzell in his other knee. A second later, Coffelt, although mortally wounded, propped himself up against the guard booth, fired his revolver at Torresola killing him instantly. Officer Coffelt was taken to the hospital and died a few hours later. The gunfight lasted less than a minute.

Collazo would live and was sentenced to death which Truman later commuted to a life sentence. Birdzell and Downs would recover from their wounds and return to duty.

As news of the attempted assassination faded from the news, another event soon grabbed the public's attention.

On November 25, the "storm of the century" hit the eastern part of the United States. The storm formed over North Carolina and moved north striking western Pennsylvania, eastern Ohio and West Virginia. The storm blanketed these areas with several feet of snow making travel impossible in some places for almost a week.

At the same time, a major windstorm covered a far greater area going up into the New England area. This storm was of hurricane-like force. It killed 353 people and injured over 160. It impacted twenty-two states and caused over $66 million in damages. The storm ended on November 30.

In 1950, most Americans received their news from daily papers, printed publications and the radio. Television coverage of news events was still years away and the immediacy of live, visual reporting had not yet come into living rooms.

So while the Marines were moving north into the Chosin Reservoir area, the news media was focused on the security of the president and the hardships and tragedy of the ongoing storm. People in America were dealing with their own winter worries. But soon the Korean War and the Marines would be back on every front page.

9

THE CHINESE ATTACK

November 27

HUNGNAM

On the morning of November 27, not hearing word to the contrary from General Shepherd, Eddie Craig departed on emergency leave for the States, to be with his dying father. Smith's right hand man was no longer available to him, but Smith who lost his own father when he was seven, understood why Craig had to go home.

Smith had been ordered to launch his attack on the twenty-seventh. While the 1ˢᵗ Marine Division had over 20,000 men, only 15,000 were currently in the Chosin area. As planned, Murray's 5ᵗʰ Marines passed through Litzenberg's 7ᵗʰ Marines and attacked west. The 5ᵗʰ Marines only advanced about 1,500 yards because of enemy resistance. General Sung had made a major strategic error. It was the first appearance of Chinese troops since the battle at Sudong on November 2. Instead of letting the 5ᵗʰ Marines pass by, the Chinese attacked too early, preventing the isolation of the 5ᵗʰ Marines from the 7ᵗʰ Marines. Had the Chinese waited one more day, the two regiments would have been separated and probably been unable to help each other. The Marines did not realize it at the time but this was a blessing because it prevented them from moving further away from Yudam-ni.

Smith spent the day at Hungnam reviewing the plans to relocate his command post north to Hagaru. Some of the division staff had already moved and Smith would make the trip the next day. Unlike MacArthur who was 700 miles away in Tokyo, Smith was with his men.

EAST SIDE OF CHOSIN

By early afternoon on the east side of the reservoir, Army Lieutenant Colonel Faith had moved his battalion north of the Pungnyuri Inlet and set up his perimeter. Faith's command post was in the center.

Faith had applied to attend West Point but was turned down for vision issues. He enlisted and rose through the ranks. Most of his service had been in administrative posts which did nothing to prepare him for the cataclysmic struggle that would soon begin.

Lieutenant Colonel Reilly, who commanded the Army's 3rd Battalion of the 31st Infantry, arrived that afternoon followed by Lieutenant Colonel Embree's 57th Field Artillery Battalion.

Colonel MacLean, commanding officer of the Army's 31st Regiment, put Embree's and Reilly's battalions into position south of Pungnyuri Inlet. MacLean sent a recon platoon in jeeps with machine guns to patrol the valley to the north. These men and their vehicles disappeared and were never seen again.

So now, the stage was being set. The Marine's 5th and 7th Regiments were regrouping on the west side of the reservoir at Yudam-ni and the Army's 31st Regiment was assuming responsibility for the east side of the reservoir. Colonel Litzenberg had sent Captain Barber's F Company of 240 men to Toktong Pass. Chesty Puller's 1st Marine Regiment was spread out along the road. His 1st Battalion was at Chinhung-ni, his 2nd Battalion was at Koto-ri and his 3rd Battalion was at Hagaru.

YUDAM-NI

The perfect storm was quickly developing and the eight Chinese divisions with 80,000 soldiers, so cleverly hidden in the hills and valleys surrounding the reservoir, were ready to spring the trap on the unsuspecting

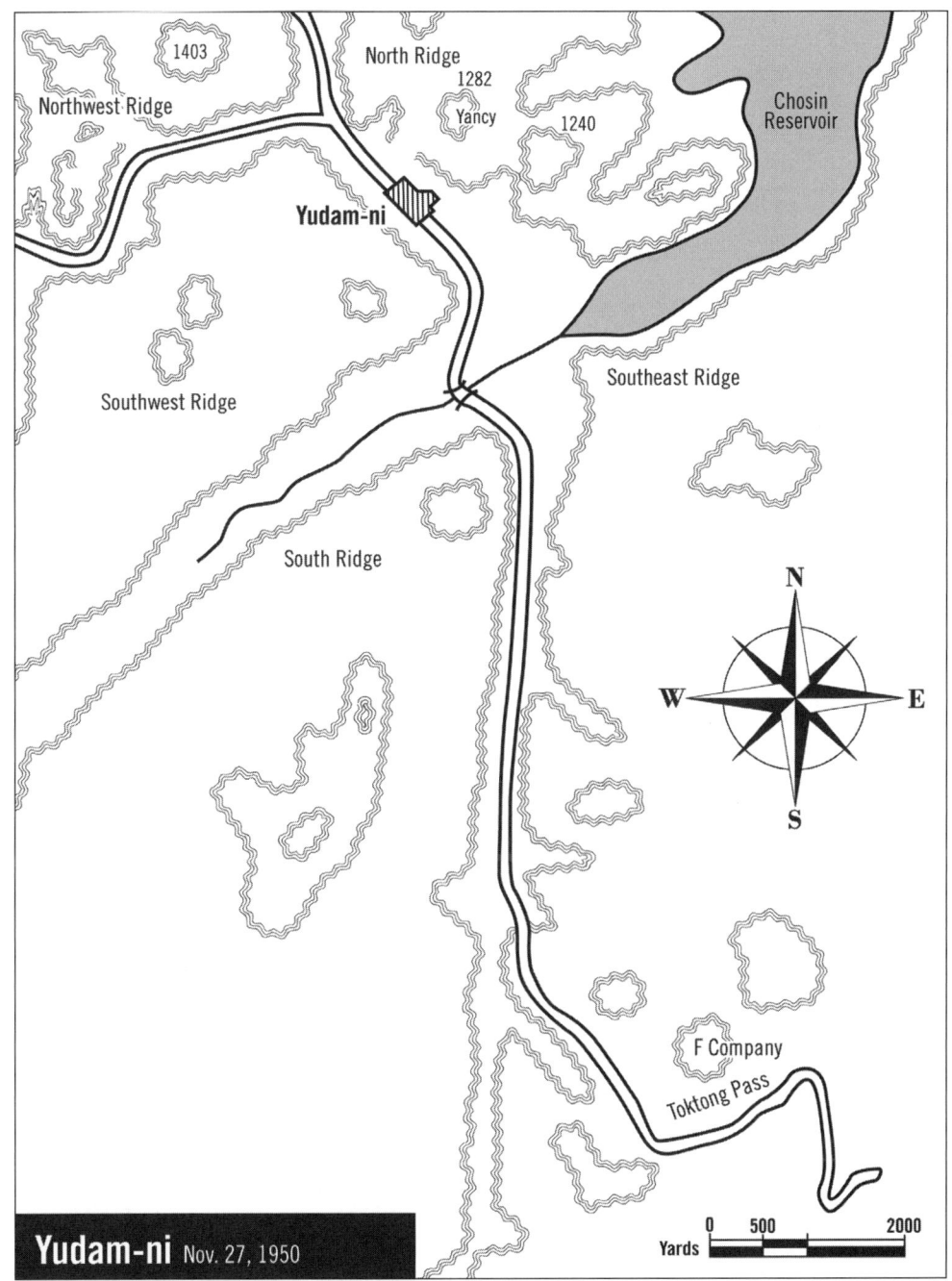

1403

North Ridge

1282

Northwest Ridge

Yancy

1240

Chosin
Reservoir

Yudam-ni

Southeast Ridge

Southwest Ridge

South Ridge

N

W   E

S

F Company

Toktong Pass

0   500   2000

Yards

Yudam-ni Nov. 27, 1950

U.S. troops. As ordered by Chairman Mao, General Sung was focusing the bulk of his troops on the 1st Marine Division.

Yudam-ni lies in the middle of a wide valley surrounded by five large ridges, named by the Marines as follows: North, Northwest, Southwest, South and Southeast. These were not fancy or clever names. They were practical ones and they were named with regard to the direction they faced from the village.

When nightfall arrived, there were a total of ten rifle companies of both Marine regiments holding the high ground around Yudam-ni. In the valley, two battalions of the 5th Marines were protecting the village. There were 8,000 Marines at Yudam-ni. Litzenberg and Murray each had their command posts in the village; Barber's F Company was isolated at Toktong Pass.

The Marines tried to dig in but the ground was frozen solid. All Marines carried an entrenching tool which they used to dig their foxholes. However, in the extreme cold this tool became virtually useless. The men were exhausted and they were numb from the cold. There was a fifty-percent alert: half of the men could sleep while the others stood watch. The night was pitch black. There was a slight wind that made it difficult to hear and it cut into exposed skin. It was twenty degrees below zero. Human beings did not belong out in these kinds of weather conditions. The Marines were not expecting a major attack but they were ready.

In this subzero weather, all vehicle engines had to run twenty-four hours a day to prevent them from freezing. Fuel was being used at an alarming rate. Hot meals would freeze before the Marine carrying it could sit down to eat.

As it grew dark, Litzenberg and Murray didn't know it yet but several divisions of Chinese had surrounded the Marines at Yudam-ni. In addition, another Chinese division was heading south to cut the road at Toktong Pass occupied by Barber's F Company, 7th Marines. And on the east side of Chosin Reservoir, two Chinese divisions were surrounding the Army units there.

Shortly after 9:00 pm, thousands of hardened Chinese communist soldiers came storming out of the cold, dark night and attacked the Marines at Yudam-ni. The Chinese would employ the same tactics that they had used for a thousand years in countless battles. The quiet of night suddenly exploded in blaring bugles, crashing cymbals and flares signaling their units to surge forward. It was an ancient but effective tactic that had a demoralizing psychological effect on Chinese foes. The bugles produced an eerie sound in the frigid mountain air. The Chinese hit the Marines with grenades followed by machine gun fire. The light from the armaments lent a surreal aura to the white, frozen landscape. Waves of fifty to a hundred men attacked the companies of Marines holding the ridges around Yudam-ni. When the Chinese were cut down by the deadly fire of the Marines, more would follow. Finally, the unending numbers of Chinese were able to break through and they savaged the Marines on the Northwest and North ridges.

On the Northwest Ridge, Captain Leroy Cooke's H Company, 7th Marines held the high ground on Hill 1403. When part of his right flank was pushed back, Captain Cooke led a counter attack but was killed by machine gun fire. By 3:00 am only one H Company officer was still standing. The remnants of H Company were ordered to withdraw to the rear leaving Hill 1403 in the hands of the Chinese.

More Chinese attacked the North Ridge which was held by two widely separated companies of the 7th Marines: E Company commanded by Captain Walter Phillips on Hill 1282 and D Company, commanded by Captain Milton Hull on Hill 1240.

One of Captain Phillip's platoon leaders was 1st Lieutenant John Yancey. During World War II, he had served as a corporal with Carlson's Raiders, an elite Marine unit serving in the Pacific. He was a "mustang," a former enlisted man, who became an officer. Yancey was awarded the

Navy Cross and received a battlefield commission. After World War II, he went back to his home in Little Rock, Arkansas, got married, and opened a liquor store, but he stayed in the Marine Reserves.

When the Korean War started, Yancey was thirty-two years old and when the reserves were called up he was sent to Korea with the 7th Marines. Yancey was commanding the 1st Platoon of E Company.

At midnight, the Chinese attacked in an endless mass of white-robed infantry. Only their weapons and a small bit of their faces stood out in the snow. Yancey moved from hole to hole motivating his men and distributing ammunition. A grenade fragment had pierced the bridge of his nose and he had to keep spitting the blood out of his mouth as it trickled down his throat.

The Marines on Hill 1282 tried to hold but the ranks thinned as bullets and grenades fell like rain and more Marines were killed or wounded. More would have died but the weather froze many of the Chinese grenades and they failed to detonate. John Yancey was wounded a second time when a piece of metal from another grenade blew a hole in the roof of his mouth.

Now there were nine men left in his platoon who could still fight with Yancey. Spitting blood, cursing and yelling the Marine Raider battle cry, "Gung Ho", he and his few men fought viciously to hold the hill. "Sha" or "kill" responded the attacking Chinese.

Yancey was knocked over when a Chinese soldier fired a machine gun directly at his face. One round hit him under his right eye popping his eye out of his socket. The impact sent him flying back into the snow. Lying on the ground, he saw the soldier reloading and he reached for his carbine but he couldn't find it, so he pulled his .45 pistol from his shoulder holster and fired twice at the Chinese soldier killing him. Then he carefully put his eye back in its socket.

1st Lieutenant John Yancey was finished. He was led down the hill to a medical tent. When he got there a Corpsman tied him sitting up to a tent pole so he wouldn't choke on his own blood. Amazingly this heroic Arkansas storekeeper would live.

Captain Phillips was killed late in the evening and when his executive officer First Lieutenant Raymond Ball took command, he was wounded several times and died in the battalion aid station. E Company was so badly shot up that only a third of its men were left. The ground on the hill itself was pulverized and most of the trees were shredded from the vicious assault. By daylight, the Chinese held the crest of Hill 1282.

A thousand yards to the east of Hill 1282, Captain Hull's D Company on Hill 1240, was also fighting for their lives. Their machine guns glowed from constant use. The sounds of gunfire echoed in the deep valley and the fighting on the next hill created a light show in the frozen landscape. Hull was wounded twice during the night, but he continued to lead his Marines. When the sun came up, he only had sixteen men left.

Other Marine units surrounding the Yudam-ni village came under attack that night including the command posts of the 5th and 7th Marines. No one was spared. At Yudam-ni, casualties were estimated at about 400.

Vastly outnumbered, the Marines at Yudam-ni held. Leadership and extraordinary courage allowed these men to survive the attack of thousands of Chinese soldiers and the worst Korean winter in 100 years.

Toktong Pass

Captain Bill Barber, the commanding officer of F Company, was another "mustang." He was born in Kentucky in 1919 and enlisted in the Marine Corps in 1940. Barber went through paratrooper training and stayed on as an instructor. He was so skilled at shooting that he was made a rifle instructor. He became an officer in 1943 and served with the 26th Marines at Iwo Jima as a platoon leader where he was wounded. Although he was hospitalized, he refused to stay out of action. He returned to take command of a company and was awarded a Silver Star and Purple Heart. Barber was no stranger to combat. He knew how to fight.

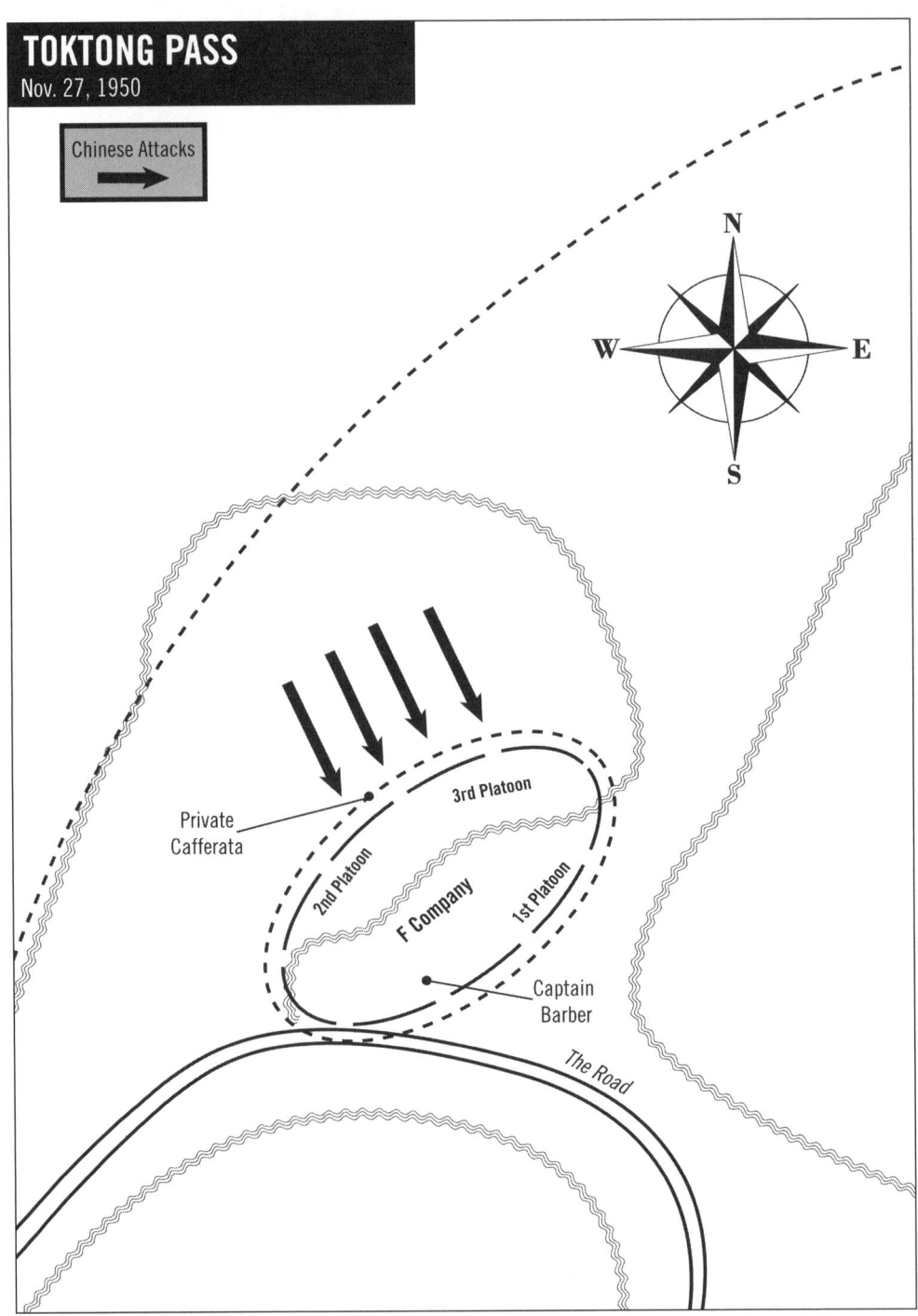

TOKTONG PASS
Nov. 27, 1950

Chinese Attacks

N
W   E
S

3rd Platoon

Private
Cafferata

2nd Platoon

F Company

1st Platoon

Captain
Barber

The Road

The Savior

When Barber first joined F Company, he was dismayed to find his men ragged and undisciplined. The company was a mix of reservists and regulars who needed to be molded into an infantry rifle company. He ordered the company to clean themselves up and shave. He taught them to look and act like Marines.

Barber's orders were to hold Toktong Pass, a 4,000 foot high pass that connected the 8,000 Marines to the north with the division command post fourteen miles south. If Barber couldn't hold the pass, those 8,000 Marines would be cut off and surrounded by the enemy. Every Marine at Toktong Pass knew why he was there and what his job was.

When his rifle company arrived at Toktong Pass, Barber set up a horseshoe shaped defensive perimeter on the brush covered hill with the 2nd platoon on the left, the 3rd platoon in the middle, and the 1st platoon on the right. Even though the ground was frozen solid, Barber made his men dig fighting holes. The Marines were cold and they were tired but they started to dig in. It took several hours of chipping away at the frozen ground but finally their foxholes were ready. They would use the cardboard boxes from C rations in the bottom of their fox holes for a little extra insulation.

At the bottom of the hill by the road, Barber placed his command tent and a first aid tent. At 9:00 pm, all 240 of his men were in place. It was twenty-five below zero and a light snow was falling.

In most closely fought battles, there is often a defining moment when one man rises above himself and does something so incredible that it significantly impacts the outcome. Many men fought heroically that night. Some of those men died and some lived. One of the lucky ones was Private Hector Cafferata.

Private Cafferata and PFC Robert Benson of the 2nd platoon were placed in a listening post thirty yards in front of the dividing line between the second and the third platoons. Cafferata and Benson were friends and had enlisted together.

Cafferata was easily the biggest man in F Company despite his young nineteen years. He stood an imposing six foot four inches tall and weighed 230 pounds. He had the look of a professional boxer but displayed a great sense of humor which often times landed him in trouble with his superiors. He grew up in rural New Jersey spending much of his time hunting and fishing. He was an excellent shot.

The Marines were on a fifty percent alert; Benson slept while Cafferata watched. Cafferata had removed his boots and had put his stocking feet in his sleeping bag for warmth. The Marines had found their feet were more likely to freeze confined in the shoe pac boots than when free in their sleeping bags.

At 2:30 am, Cafferata detected movement to his front and within seconds scores of white clad Chinese rushed past them. Cafferata woke Benson and started shooting point blank into their charging lines. While Benson struggled to put on his boots, Cafferata shot six Chinese soldiers with his first clip of eight rounds.

Cafferata and Benson grabbed all their ammo and sprinted back to a slit trench occupied by two other Marines, PFC Pomers and PFC Gerald Smith. Cafferata and the three Marines, standing shoulder to shoulder, fired into the enemy lines.

Suddenly Chinese grenades were flying in from all directions at the surrounded men. Cafferata caught two grenades in midair and hurled them back at the attackers. Then he used his entrenching shovel like a baseball bat knocking another grenade back toward the enemy. Still another grenade landed to the back of their trench. Cafferata grabbed the grenade and while throwing it back, it exploded, shredding one of his fingers.

Benson grabbed yet another Chinese grenade which exploded as he threw it. The grenade shattered his glasses and filled his eyes with debris temporarily blinding him. Unable to see through his bleeding eyes, Benson dropped down and began reloading rifles which he passed to Cafferata. The countless hours of Marine Corps weapons training made the action a reflex, needing no sight.

Pomers was blown to the other side of the trench by a concussion grenade and was knocked out. The fighting went on all night and the failure of the Chinese to break through the Marine lines at the juncture of the second and third platoons was due to the bravery and determination of these four men.

At dawn, the Chinese broke off the attack. Cafferata realized he had fought all night in just his stocking feet. He tried to retrieve his boots from his original listening post but was shot by a Chinese sniper. The bullet went through his right shoulder bounced off a rib and punctured his lung. Cafferata would live but his heroic fight was over. Shoeless, he was taken down to the medical tent at the bottom of the hill.

By morning, twenty of Barber's 240 Marines had been killed and fifty-four wounded. Some of the wounded died because the blood plasma they desperately needed had frozen and there was no way to thaw it. But despite his losses and the overwhelming odds against him, Barber and his men held Toktong Pass. It was estimated that F Company had mowed down over 450 Chinese soldiers of the estimated 1,400 in the attack. The Chinese had broken the Marine perimeter once, but they failed to take advantage, allowing the Leatherneck lines to reform.

After the dead had been placed outside the first aid tent and the wounded had been cared for, Captain Barber walked along the perimeter to check on his men. When he arrived where Cafferata and his buddies had fought all night long, he was stunned at what he saw: over 100 Chinese bodies littered the snow covered ground. Barber put his exhausted men to work stacking frozen bodies of the Chinese soldiers like cordwood to help form barricades. The sight of these piled corpses was unsettling to the battle hardened Marines. They hoped that the sight would be even more unsettling to the enemy.

The Marines at Hagaru and Koto-ri were not attacked that night but their turn would come.

EAST SIDE OF CHOSIN

On the east side of the reservoir at midnight, a massive Chinese force launched a full scale attack against the Army's 31st Regiment. Non-existent communications between the Army and Marine forces prevented the 31st Regiment from learning about the attacks against the Marines at Yudam-ni and the heavy Chinese presence in the area. At the same time, the Marines did not know their Army brethren to the east were also being attacked.

North of Pungnyuri Inlet, Lieutenant Colonel Faith's A Company came under heavy fire. His company commander was killed. Marine Captain Ed Stamford, the forward air controller, took temporary command. Soon the Chinese had the rest of the Army battalion surrounded.

South of the inlet, Lieutenant Colonel Reilly's 3rd Battalion of the 31st Infantry and the two firing batteries of Embree's 57th Field Artillery were in big trouble. The Chinese attacked these Army positions from the east and overran Reilly's command post and both artillery batteries. Reilly was wounded. To the south, Embree's artillery headquarters was also hit and he was wounded. The battle continued non-stop all night long and finally started to subside as snow began falling. At dawn, a wonderful sight greeted the Army soldiers as four Marine Corsairs, called in by Captain Stamford, worked over the Chinese with napalm, rockets and strafing fire. The Army's 31st Regiment had suffered over 200 casualties.

The Truman administration, the Joint Chiefs of Staff and MacArthur and his staff in Tokyo were totally surprised by the news of what was happening at the Chosin Reservoir. Communist China had entered the Korean War with a vengeance. Eight Chinese divisions with over 80,000 troops, had surrounded the Marines at Yudam-ni, Toktong Pass, Hagaru,

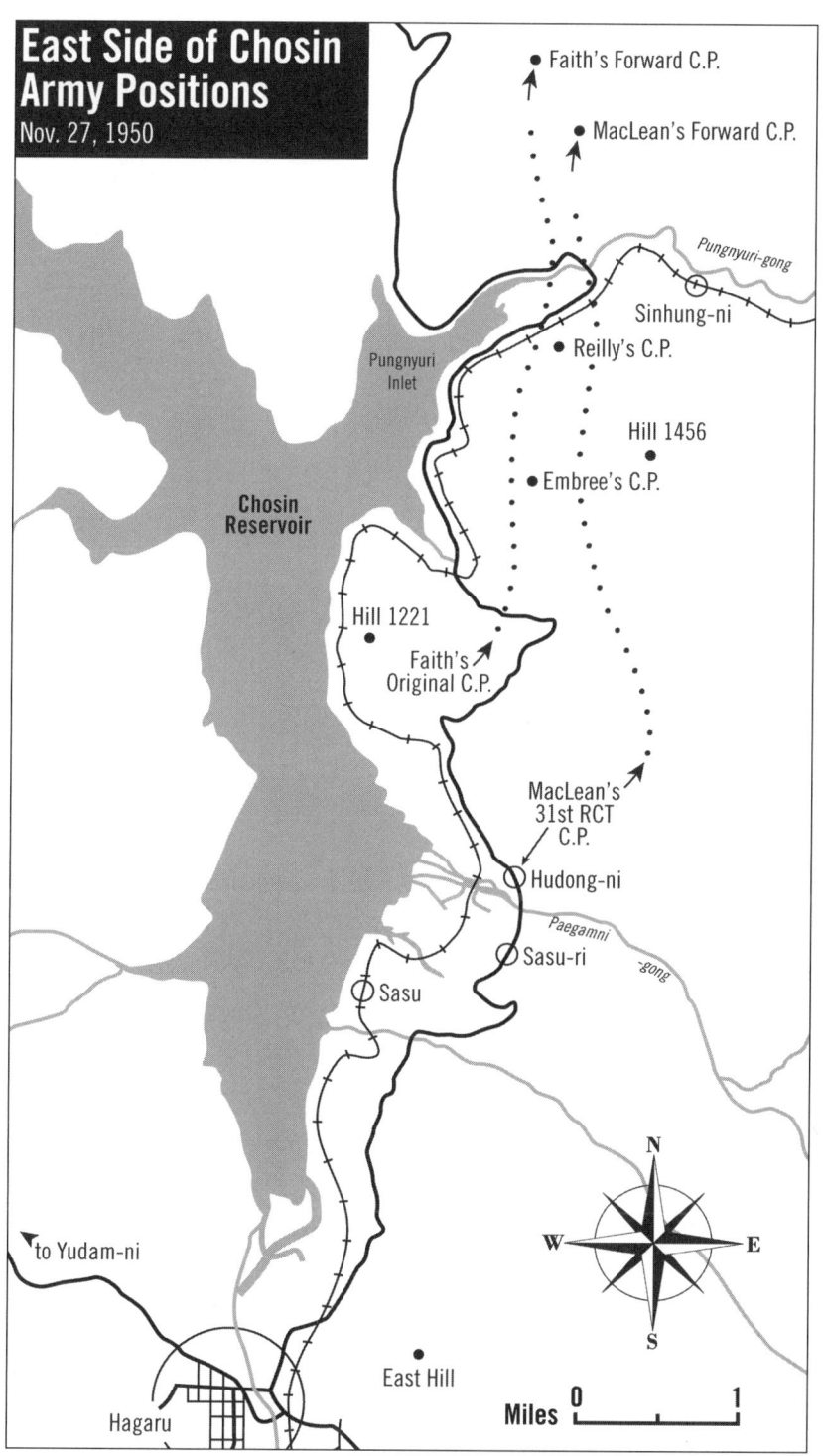

East Side of Chosin Army Positions
Nov. 27, 1950

Faith's Forward C.P.

MacLean's Forward C.P.

Pungnyuri-gong

Sinhung-ni

Pungnyuri Inlet

Reilly's C.P.

Hill 1456

Embree's C.P.

Chosin Reservoir

Hill 1221

Faith's Original C.P.

MacLean's 31st RCT C.P.

Hudong-ni

Paegamni -gong

Sasu-ri

Sasu

N

W E

S

to Yudam-ni

East Hill

Miles 0 1

Hagaru

Koto-ri and the Army on the east side of the reservoir. Everywhere along the main supply route, the road was cut, as Smith had feared. Resupply by road was now impossible. The 5th and 7th Marine Regiments at Yudam-ni, F Company at Toktong Pass and the Army Battalions on the east side of the reservoir had suffered the most and the fighting had just begun. And the cold was only getting worse.

10

THE DEFENSE OF HAGARU

November 28

YUDAM-NI

When a late fall sun came over the mountains in the morning, Murray met with Litzenberg at the 7th Marines command post. They both agreed that the only option they had was to switch from the offensive to defense. Murray canceled the attacks that were planned for the morning. But without orders from X Corps, they couldn't do anything about withdrawing from Yudam-ni.

Besides being vastly outnumbered by the Chinese, the biggest problem facing the Marines at Yudam-ni was taking care of the many wounded. A few of the lucky Marines were flown out on helicopters. Unfortunately, helicopters at that time could only carry two casualties per trip and there were hundreds of wounded men. The remaining wounded were cared for by the battalion aid stations which were soon overwhelmed. Casualties were laid out on the floor like sardines. The moans and cries of these wounded and dying men was heartbreaking. Throughout the day airdrops of critically needed medical supplies and ammunition were made but little could be done to evacuate the wounded.

HAGARU

At 11:00 am, as planned, General Smith and key members of his staff flew north into Hagaru by helicopter to open his command post. Since the road was cut off by the Chinese, there was no other way to get there. Personnel of the division headquarters had been arriving in Hagaru since November 24. After Smith landed, he was driven to his command post. It was located at the northern edge of the village near the slope of a pine covered hill. When he climbed out of his jeep, he stopped for a second and looked around at the expanse of the base. Most of the existing structures had thatched roofs but there were a few multi-storied wooden buildings. He had seen pictures, as a kid, of the Yukon Gold Rush of 1898. Hagaru looked a lot like one of those bleak mining camps: snow covered tents, frozen roads bustling with activity and cold, miserable men.

He was quartered in a frame house constructed by the Japanese during their occupation of Korea. The remainder of the command post personnel were housed in a complex of tents and vans. Hagaru was just a small mountain road town that would quickly be filled up as it became the key headquarters and supply base. Smith's house was surrounded by a fence. Smith thought that the house might have had a little bit of charm if it had been any place other than Hagaru. In the house, there were three rooms and a kitchen. This was a place for Smith to only sleep and eat. His quarters were sparsely furnished and his bed was a canvas cot. Smith knew he would be spending the majority of his time in the command tent. Because the road was closed, Smith only had the few personal items he brought with him in the helicopter. The remainder of his gear stayed at Hungnam.

When Smith entered his new quarters he saw a picture of the Soviet dictator Joseph Stalin hanging on the wall. A staff officer started to take it down, but the general said, "Leave it there; it might inspire us."

Smith knew that the 5th and 7th Marines had been attacked at Yudam-ni but he didn't know just how bad it had been. Communication was difficult because of weak radio signals over the mountainous terrain. The extreme cold played havoc with the radios and their batteries. Helicopters had to continually fly in replacement batteries for those that ran out quickly in the subzero weather.

Shortly after Smith's arrival, the X Corps Commander, Major General Almond and his aide, First Lieutenant Alexander Haig landed in a single engine plane at the partially completed airstrip in Hagaru. Almond and his aide were driven to Smith's command post.

"General Smith, what happened at Yudam-ni?" asked Almond.

"We don't have all the details yet, General, but Murray and Litzenberg have come under serious attack. Casualties are piling up quickly. There are a lot of Chinese out there; we just don't know how many," replied Smith.

"Where did they all come from?" asked Almond.

"China, I would imagine," responded Smith. His tolerance for Almond was almost nonexistent.

Almond chose to overlook the impudent response and opted not to engage Smith in a debate over the Chinese presence. He still believed that MacArthur's intelligence should be relied upon and despite clear evidence to the contrary thought the race to the Yalu could continue.

"General Smith, I need to borrow one of your helicopters; I want to talk to Colonel MacLean and Lieutenant Colonel Faith and find out what's going on over on the east side. As you know, we have had real communication problems because of these damn mountains."

"General, when you get back to the landing strip, I'll have a helicopter waiting for you."

With that, Almond and Haig departed for the strip. Smith breathed a sigh of relief and returned to his work.

EAST SIDE OF CHOSIN

Colonel MacLean, who had spent the night of November 27 at Faith's command post, left at dawn that morning to return to his own advanced command post, a short distance south. Twenty minutes later, Almond's helicopter sat down on the east side at MacLean's forward command post. There, Almond and MacLean talked for a while and then they jeeped forward to Faith's command post to confer with him.

General Almond knew little about the entire situation on the east side of the reservoir. Communications remained terrible. The general was greeted at the jeep by Faith, who accompanied the visitors to his command post.

"Colonel Faith, I'd like a full report on the fighting last night."

"General," said Faith quickly, "my battalion was attacked all night by elements of two Chinese divisions. It was a rough night. We had no intelligence to alert us of the size of the Chinese force. We were severely outnumbered and took significant casualties. Some officers have been lost."

Almond, looking so out of place next to Faith's weary soldiers, was dressed in a new winter parka and freshly starched and pressed snow pants. Some of Faith's men were still in summer uniforms. With a dismissing wave of his hand, Almond lectured, "That's impossible. There aren't two Chinese Communist Divisions in the whole of North Korea. The enemy who is delaying you for the moment is nothing more than a remnant of a Chinese division fleeing north. We are still attacking and we are going all the way to the Yalu. For God's sake man, don't let a bunch of Chinese laundrymen stop you."

Faith looked at the X Corps commander in disbelief. There were Chinese bodies surrounding his perimeter. Didn't General Almond see these? But Faith and everyone else at this meeting had no idea of just how bad the situation was at the inlet with Reilly's 3rd Battalion and Embree's 57th Artillery Battalion.

"Colonel Faith," said General Almond, "I have several Silver Stars to award to you and some of your men for the outstanding work last night. Select two of your men to receive the Silver Star."

"General Almond," protested Faith, "there were far more than three men who performed heroically last night. A lot of men are more deserving than I am."

"No time for modesty, Colonel," interrupted Almond. "Now who are the other two?"

Just at that time, Lieutenant Smalley who had been wounded during the night walked by.

"Lieutenant Smalley, come over here," Faith said. He also saw Sergeant Stanley, a mess sergeant, who performed admirably in the fight.

"Sergeant Stanley, join us."

Faith introduced both men and continued, "General Almond has a Silver Star to award you for your outstanding service in last night's battle."

"Colonel Faith, Lieutenant Smalley, Sergeant Stanley," started Almond. "It is my privilege to award the Silver Star for gallantry in the battle last night, November 27, 1950. The Army and your country are proud of you."

With this, the Commander of X Corps pinned a Silver Star on each of their jackets.

The recipients exchanged salutes with General Almond and then he turned to leave the area.

Something welled up inside Faith. Maybe it was the whole idea of being isolated in this bitterly cold place and under fire; maybe it was the foolish rush to attack north even though they were under staffed and under supplied. And maybe it was that Almond just didn't get it. But something exploded. As Almond walked away, Faith ripped the medal from his jacket and threw it in the snow. He thought of his father, a brigadier general in the Army, and the way he had been brought up to be an honorable man.

"What a damn farce," Faith mumbled. His anger and disgust no longer contained.

Lieutenant Smalley followed suit ripping off his medal as well. Sergeant Stanley removed his award and put it in his jacket pocket.

Due to abysmal communications between his units, MacLean was also completely unaware of the pounding that his 3rd Battalion had taken at the inlet just four miles south.

Sadly, neither officer knew just how weak and vulnerable MacLean's task force was. Just defending itself would be a monumental task let alone launching an offensive attack north. Continued delusional thinking would wreak devastating results.

General Almond returned to Hagaru and from there flew back to his command post at Hungnam. General Barr, the CO of the Army's 7th Infantry Division, and Almond conferred and both felt that the push north should happen as soon as the Army's 2nd Battalion joined the 31st Regiment east of the reservoir. Later that day, Almond received orders from MacArthur to report to his office in Tokyo, immediately.

At the same time north of Hagaru at Hudong-ni, Army Brigadier General Hodes met with the Tank Commander Captain Bob Drake.

"Bob," said General Hodes, "we need to get these tanks north to support the infantry at the inlet. We're too strung out. I'll ride along with you."

They started in Drake's jeep accompanied by sixteen tanks. The fourth platoon of tanks remained to defend the supply dumps at Hudong-ni.

When they neared Hill 1221, they immediately came under heavy Chinese fire. Captain Drake quickly divided his force into three platoons and started to assault the hill.

Icy slopes, frozen roads, and heavy enemy fire repeatedly stopped Drake's tanks on Hill 1221. General Hodes and Captain Drake conferred and agreed that infantry support was needed if they were to take the hill. Drake would try again to dislodge the Chinese with air strikes and infantry.

General Hodes took Drake's jeep and drove back to Hudong-ni, arriving about noon. Hodes knew the only possible assistance would be found at Hagaru and he planned to get a jeep for the five mile drive. A suggestion was made that for General Hodes' safety, he should make the trip to Hagaru in a tank. Hodes reluctantly agreed and took the slower tank ride south to Hagaru arriving in early afternoon without incident.

Hodes found General Smith in the staff tent and immediately explained his predicament. He said to Smith, "General, I've just returned from the east side of the reservoir where the 31st Regiment was hit hard last night. We tried to attack north with a company of tanks to help the 31st but we were stopped at Hill 1221. We're going to need some help over there."

"General Hodes," said Smith, "I'm not sure that you're aware of this, but the 5th and 7th Marines were brutally attacked last night up at Yudam-ni. We can't do anything to assist you until we help them out. There has obviously been a lot of fighting both up on the east side and up at Yudam-ni and I expect that Hagaru will be next. We are stretched pretty thin here at Hagaru. I only have two rifle companies and some service troops. General, I just arrived here a few hours ago from Hungnam and I'm still trying to get a handle on this. My biggest concern now is holding this place. If we lose it, the 5th and 7th are done for as is the 31st Regiment."

Smith and Hodes spoke for several minutes more and then Hodes left Smith's command post. Hodes was desperate for help but understood why Smith could not provide it.

Late in the afternoon, Colonel Bowser, Smith's operation officer, flew by helicopter from Hungnam to Hagaru. When he arrived, he reported immediately to General Smith. From this point forward, Smith and Bowser would be joined at the hip. With Eddie Craig back in the states, the general would have to rely on his operations officer even more. And Al Bowser was the perfect man to rely on. He was a careful, inventive planner who did his job well.

Smith filled Bowser in on the fighting at Yudam-ni, Toktong Pass and the east side of the reservoir, as well as General Hodes' situation and request for help.

"Well General, they can't say you didn't warn everyone of this possibility," said Bowser.

"I take no satisfaction in being right on this one," said Smith. "Al, it looks like it will be our turn tonight. We've got to be ready."

"Sir," started Bowser, "on the way up here I counted nine unmanned road blocks and saw quite a few abandoned American vehicles."

"Al," Smith continued, "it's obvious that the road between Koto-ri and Hagaru has been cut. The same goes for the road between here and Yudam-ni. Issue orders to have Litzenberg's 7th Marines open the road between here and Yudam-ni. I can't order the 5th and 7th to withdraw until we get the order from X Corps, but we need to get that exit route open. Also, tell Murray's 5th Marines to hold up the attack north. Have Puller's 1st Marines clear the road between Koto-ri and here. Have you appointed a base commander for Hagaru yet?"

"Yes, sir. It's Lieutenant Colonel Ridge, Commanding Officer 3rd Battalion 1st Marines, one of Puller's men," replied Bowser.

"Good choice," said the general.

After the meeting between Bowser and Smith ended, preparations for the defense of Hagaru went into overdrive. Colonel Ridge spent the remaining three daylight hours tightening up the defensive perimeter around the Marine base at Hagaru. He seemed to be everywhere at once as he oversaw gun placements and unit assignments.

A day earlier on the morning of November 27, Ridge on his own initiative had already started preparations for the defense of Hagaru. Based on recently received intelligence, Ridge was convinced the terrain around Hagaru offered the Chinese two likely avenues of approach. One was a draw leading toward the southwest portion of the defensive perimeter near the construction site of the airfield. Here he would place his only two rifle companies, H and I. The other likely spot was a hill to the east of town which the Marines simply named East Hill.

H Company was commanded by Captain Clarence Corley Jr. Corley graduated from Louisiana College where he was captain of the football team. He served in World War II in the South Pacific.

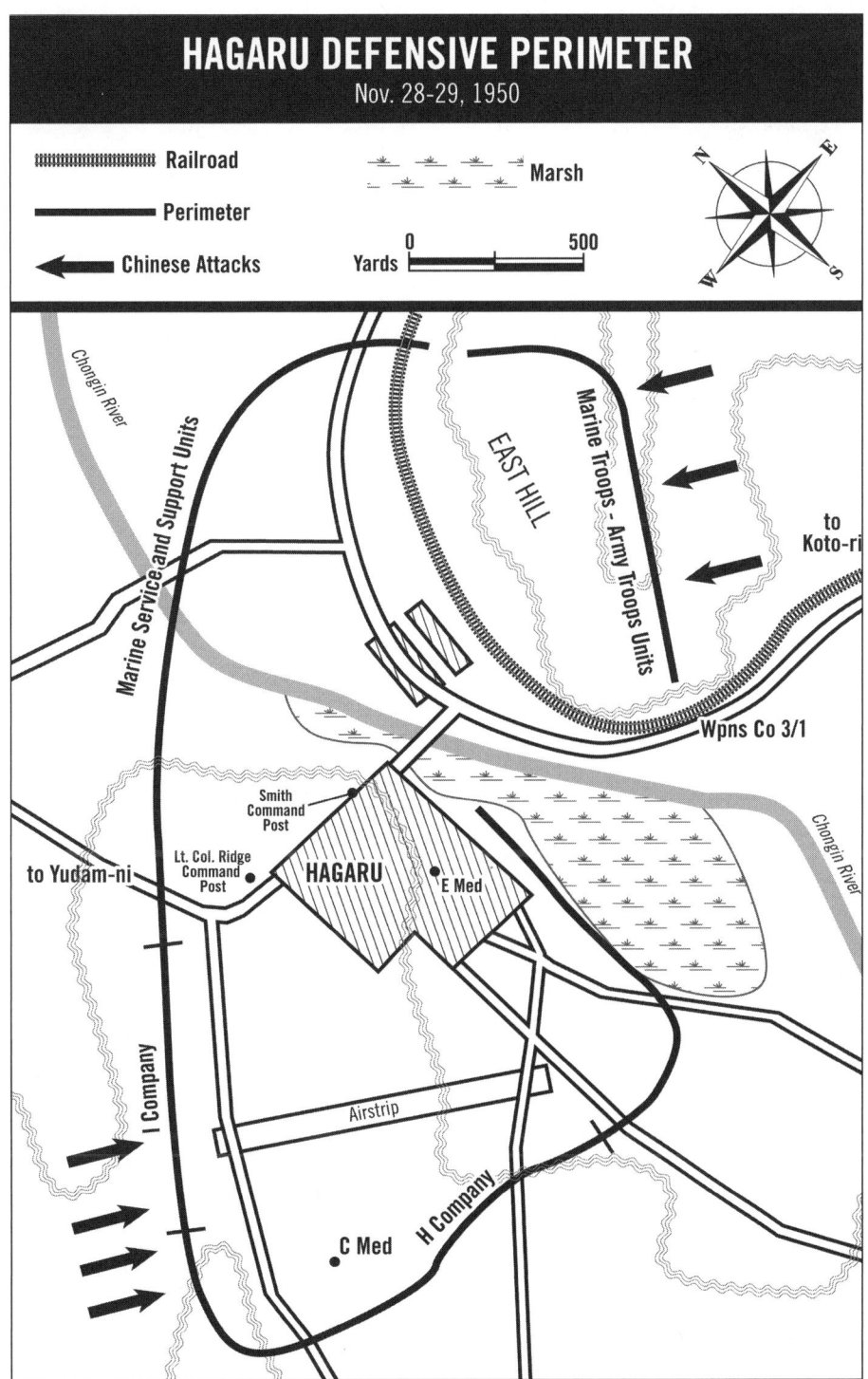

HAGARU DEFENSIVE PERIMETER
Nov. 28-29, 1950

Railroad

Perimeter

Chinese Attacks

Marsh

Yards 0 — 500

Chongin River

Marine Service and Support Units

EAST HILL

Marine Troops - Army Troops Units

to Koto-ri

Wpns Co 3/1

Chongin River

Smith Command Post

Lt. Col. Ridge Command Post

to Yudam-ni

HAGARU

E Med

I Company

Airstrip

H Company

C Med

I Company was commanded by First Lieutenant Joseph Fisher. He was given the nickname "Bull" both because of his aggressive leadership style and his stature. He stood six feet two inches and weighed 235 pounds. Fisher was a platoon sergeant at Iwo Jima and was later commissioned an officer.

Both Corley and Fisher put each of their three infantry platoons on line. That left no reserve. Three tanks were placed between H and I companies and they would cover a large draw through which the Chinese were expected to come.

Lieutenant Fisher had scrounged C-3 plastic explosive and 1,000 sandbags from the engineers working on the airstrip. Because the ground was frozen solid, the C-3 was placed in empty ration cans and used to blast through the thick, icy crust. The frozen chunks of earth were then used to fill the empty sandbags which were placed in front of their fighting holes. A limited supply of concertina and other barbed wire was strung out along their front. Tin ration cans filled with rocks were hung from the low wires to serve as noisemakers. The Marines worked furiously to complete these preparations.

Next, Fisher and Corley set out five gallon cans filled with gasoline and rigged them with white phosphorus grenades to which strings were tied and when the strings were pulled, the grenades would explode creating a flaming wall as the gasoline ignited. Trip flares and land mines were strategically placed along their front.

As they watched the men setting these traps, Fisher told Corley, "We'll give the Chinese the American version of fireworks."

Corley agreed that these preparations would light up the night when the Chinese attacked. Snow was falling and the temperature dropping but the Marines preparing for the expected attack were full of adrenaline. They did not expect to get much sleep.

East Hill was a steep, rocky, ice covered ridge that overlooked the town of Hagaru. During the day Ridge did not get this hill sufficiently manned, so at 5:00 pm he sent up additional Marine and Army non-combat units to protect this strategic location.

On the right side of the hill were the Army engineers. In order to give this outfit some infantry expertise, Marine Captain John Shelnutt, Ridges' weapons company executive officer was sent to command this Army unit. He took with him his radio operator, PFC Bruno Podolak.

On the left side, Lieutenant Colonel Ridge placed his Marine unit made up of service troops: cooks, clerks and band members, it didn't matter. They were still fighting men. They were Marines. Ridge ordered these two units to meet on the crest of the hill. They came from two different directions climbing the hill in the dark. They made it to the top in about an hour but for some unknown reason, maybe it was the complete darkness, the hook-up failed after several attempts.

Shortly after dark, General Smith received a call on the radio from Colonel Litzenberg.

"Sir, we can't open the road. The Chinese hold the high ground and we started taking too many casualties so I ordered my men back to Yudam-ni. The enemy is well positioned in bunkers and we need to dislodge them."

"Litz, you try again in the morning. Do what you can to call in air support. We have to open that road. You know it's your only way out."

Fifteen minutes later, Puller called in with the same story. "The Chinese controlled the road, the fighting was intense and because night was coming on we pulled back."

Smith could tell instantly that Puller was not happy to have to make this report.

"Lewie," said Smith, "you have to force that road open."

"Sir, more troops arrived at Koto-ri this afternoon: Colonel Ridge's G Company, the Army's B Company of the 31st Infantry and Lieutenant Colonel Drysdale's Royal Marines. I'm putting Lieutenant Colonel Drysdale in command of all three outfits and I've ordered him to fight his way to Hagaru first thing in the morning. I'm calling this unit 'Task Force Drysdale.' We'll get him air support once we have daylight."

"Sounds like a good plan. Those Brits have a great reputation. Keep me informed Lewie."

"Yes sir," replied Puller.

Smith looked at Bowser, "We're not going to get any help from Yudam-ni or Koto-ri tonight. Have Colonel Ridge come over here when he can."

A half an hour later, Ridge arrived. Smith motioned Ridge to sit down at a chair in front of a table Smith was using for a desk.

"Colonel, I think we're going to have some visitors tonight and I just want to make sure we're ready."

"Sir, yesterday Lieutenant Carey, the Intelligence Officer for the battalion, had recruited a number of Koreans from the area to serve as counter intelligence agents or 'line-crossers.' Carey had these men collect information from all the civilians that live in the area. They all had the same story to tell. Chinese soldiers had kicked them out of their houses and it was clear that there were a lot of Chinese in the area."

"Did Lieutenant Carey have an estimate of how many Chinese?" asked Smith.

"Carey thought the numbers he was getting might be exaggerated so this morning he sent out two of his line-crossers to make direct contact with some Chinese soldiers. They were to leave the perimeter and go up into the hills and find out whatever they could."

"What did he find out?" asked Smith.

"The agents returned early this afternoon and brought back some information he thinks is pretty reliable. They indicated that the Chinese are gathering to the south and west of Hagaru and that they are well led and well-armed."

"How far away are they?" asked Smith.

"About five miles."

"Do we know how many?"

"Sir, it's estimated that the enemy is in division strength and that they are going to attack tonight. It gets dark at five and since they are five miles from here, it might take four and a half hours to get here. I would expect an attack around 9:30 pm."

Smith asked, "Does Lieutenant Carey think these line-crossers are absolutely trustworthy?"

"Yes, he does sir."

"Are we ready?"

"We are sir," replied Ridge. "We have about 3,300 Marines most of which are service and support units. So going in we are outnumbered about three to one. I only have my two rifle companies, H and I. We have the Marine engineers who are working on the airstrip. They will continue working unless things really get bad and then I will have to pull them off. In addition, we have 500 Army troops. I have Marine service and support units on the northern part of the perimeter. On East Hill, I put a combination of Marine and Army service units. I think that if we do get hit, and I think it's almost a sure thing, they are going to hit us hard along the southern end of the perimeter where we are building the airstrip. All the intelligence reports from patrols, air recon and civilian feedback indicates that the enemy is moving in small groups to the southwest of Hagaru. Because of that I placed my two rifle companies there. My third rifle company, G Company, is stuck down at Koto-ri. They will be coming in when they reopen the road. As soon as they get here I'll put them on East Hill. Well, sir, that's about it."

As Ridge stood up to leave, General Smith said, "Good luck, Colonel."

"Thank you, sir," replied Ridge.

An hour later, Smith and Bowser left the command tent and went back to Smith's quarters to have a quick supper. After the mess sergeant had placed their plates on the table, the phone rang. Bowser stood up, answered it and after a short conversation, Bowser sat back down at the table.

Looking at General Smith, he said, "Sir that was Lieutenant Colonel Ridge. All the Marines and soldiers are now out on the perimeter. He said he assigned extra men to protect the supply dumps, division headquarters, medical facilities and airstrip."

Bowser continued, "Also, General, all personnel not essential to the operation of the command post were sent out to man positions on the southern part of the perimeter. There is no 'rear' at Hagaru. Tonight, we've got everyone out on the line; cooks, clerks, supply people, everyone."

Smith, with pride in his voice replied, "Al, this is when we thank God that every Marine is first a rifleman. All we can do now is wait."

After finishing their meal, Smith and Bowser returned to the command tent, armed with their pistols.

Fortunately, due to Smith's foresight, the Hagaru base had already been stockpiled with rations, fuel, ammunition and medical supplies. The Marines would not run out of supplies, but the stockpiles made Hagaru a valuable target for the Chinese.

At 10:00 pm, there was a light snow falling at Hagaru. Suddenly, bugles, whistles, exploding grenades and mortar fire shattered this peaceful silence. The Chinese crashed into the lines of H and I Companies. But the Marines were ready.

When the Chinese ran into the maze of booby traps, the mines and grenades exploded, the flaming gasoline lit up the front line and the Marines responded, opening up with rifles, machine guns, and mortars. The carnage was horrific. But more Chinese kept coming and finally punched through the center of H Company's line, creating chaos for the next several hours. Captain Corley tried to plug the hole in his lines but couldn't. The old football captain didn't have a deep bench to call on. Several Chinese soldiers actually broke through all the way to the airfield where they were shot and killed by engineers. Other Chinese who broke through the lines at Hagaru took to looting instead of pressing their advantage. Loaded with looted supplies, they were easy targets for the Marines.

At midnight, the fighting became so intense that no area in the perimeter was safe. C Medical Company, 200 yards to the rear of I Company, was hit by enemy fire as the surgeons and corpsmen worked on the wounded. Fortunately no one was hit. Several enemy rounds even penetrated the walls of General Smith's house, ricocheting off pots and pans in his galley.

At 4:30 am, Captain Corley, although wounded by a bullet in his arm, launched a counterattack pushing the Chinese back. Around the same time, Lieutenant Fisher ordered his men to set two Korean houses afire. For some strange reason, maybe it was just the warmth, the Chinese soldiers were drawn to the flames where the riflemen of I Company, with the support of machine guns from the tanks, mowed them down.

When the Chinese attacked East Hill after midnight, the situation went from bad to worse. The Chinese hit the left flank of the Army engineers, collapsing their line and causing them to fall back. Some of the soldiers fled the hill. Other soldiers remained and most of them died.

When daybreak came, the Chinese broke off the fight with H and I Companies at the southern end of the perimeter leaving behind 750 of their dead. The battle for Hagaru was not over; it was just beginning. Bodies littered the landscape, and the freezing temperatures turned them into blocks of ice.

Fourteen miles to the north, the Marines at Yudam-ni waited for the attack. Murray and Litzenberg tightened up the perimeter but the attack never came. An uneasy quiet prevailed, but after the night before, the men took whatever rest they could.

TOKTONG PASS

In the afternoon, Captain Barber was ordered by Colonel Litzenberg to fight his way to Yudam-ni but because of the many wounded and the fact that his company was surrounded, it was impossible to obey this order. He explained the situation to Litzenberg and asked for resupply by air for much needed ammo. Litzenberg wished him good luck and promised that ammo was on the way as soon as the planes could drop it.

At night a restless quiet settled over the hill until 2:15 am when the Chinese came charging down the snow covered ground, bugles blaring and attacked the Marines. The Chinese usually came in the darkest, coldest part of the night. That night, the snow was swirling, hampering visibility. The Chinese attacked through the thick veil of snow. In their white coats they looked like deranged snowmen. A Marine shot one of the buglers and the last note of the dying man lingered in the air. Barber left his command post and ran up the hill to fire up his men. When he reached the northeast corner of his perimeter, he was hit by a machine gun bullet in the upper part of his leg. He took out a handkerchief, plugged the hole and kept moving. His men knew that night why Barber had won a Silver Star at Iwo Jima.

It was utter chaos for the remaining evening hours and Barber continued to limp around bucking up his men. The Marines held. In the fighting, 200 Chinese were killed. Before the sun came up that morning, the cries of the Chinese wounded echoed through the darkness until one by one they slowly froze to death in the bitter cold. When daylight came, the captain was helped downhill to the med tent where he was treated. Five additional Marines were killed and twenty-nine more wounded. Captain Barber would continue to command, now missing over a third of his troops.

That night the Chinese did not attack Puller's Marines at Koto-ri; the road was blocked to the north but to the south at Chinhung-ni the road was still open.

EAST SIDE OF CHOSIN

Emboldened by their success the previous night, the Chinese would attempt to finish off the Army soldiers north of the inlet. The attackers did not even wait for complete darkness. The Chinese threw wave after wave of their peasant army against automatic weapons. Chinese casualties were high but they kept coming. The numbers of American wounded and killed grew. It became obvious to both MacLean and Faith that an immediate breakout to reach and join the 3rd Battalion south of the inlet was necessary or they would be annihilated.

All the Marines and soldiers in the Chosin Reservoir area were now fighting for their lives. With everyone going to the defensive, the ill-conceived race north to the Yalu was finally over. General Almond's offensive battle plan was dead in just two days.

And so, in the coming days and nights, why would the Marines hold on at Yudam-ni, Toktong Pass, Hagaru and Koto-ri, while the Army counterparts on the other side of the reservoir would not? The answer to this could be found in the memory of that brutal bastard of a drill instructor that had taught these young Marines three valuable lessons: because they were Marines, much was expected of them; they were part of an elite fighting force that dated back to 1775; and it was an unforgivable sin to let your fellow Marines down.

These hard earned lessons learned in Marine boot camp and the outstanding leadership of the officers and senior enlisted men of the 1st Marine Division would heavily influence the outcome of the battle. And so would something else called "esprit de corps."

It also helped that their commander, General Smith, had put his own career in jeopardy to do all he could to safeguard his men.

TOKYO

At 12:30 am, General Almond arrived in Tokyo for his meeting with General MacArthur. General Walton Walker, the commanding officer of the 8th Army, which was now in full retreat in the west, was also present.

MacArthur asked Walker, "Where can the 8th Army make a successful stand?"

"We can hold Pyongyang and we'll set up a defensive line north and east of the city," replied Walker.

Then MacArthur asked the same question of Almond. He responded with his usual bravado, obviously trying to one-up General Walker.

"We will continue to attack west and north and proceed to the Yalu as planned," responded Almond. Clearly he still had no idea of how bad it was in his area of operations. There was no attack to continue. His surrounded men were fighting for their lives and losing.

But MacArthur, because of the rapidly deteriorating situation of the 8th Army, had already made up his mind. He looked at Walker and said, "You hold the area around Pyongyang but withdraw as needed to prevent Chinese forces from moving around your right flank."

Then MacArthur ordered Almond, "End the offensive action, pull back and concentrate X Corps in the Hamhung-Hungnam area."

A disappointed Almond left Tokyo at noon and flew back to Korea. When he arrived, he issued orders for X Corps to stop the attack and to withdraw to Hamhung.

MacArthur had gambled and MacArthur had lost.

11

THE NEWS WAS ALL BAD

November 29

EAST SIDE OF CHOSIN

At 4:30 am, Army Colonel MacLean and Lieutenant Colonel Faith started their withdrawal moving south to the inlet. The wounded were loaded into trucks for the hazardous four mile trip.

The command party came under fire shortly before reaching the inlet. The column split into two groups, one with MacLean and the other with Faith.

At this time, Colonel MacLean saw a column of troops approaching from the south. MacLean, overjoyed at the thought that this was the long anticipated arrival of his 2nd Battalion to reinforce his troops, ran waving his arms wildly and shouting, "Those are my boys." But he was mistaken.

He ran alone out onto the ice toward the other side of the inlet. Chinese bullets riddled MacLean's body but again and again he got up and started toward the column.

MacLean was never to be seen again as Chinese troops took hold of him and dragged him into the brush.

Lieutenant Colonel Don Faith, being the senior surviving officer, assumed command of the 31st Regimental Task Force. From that point on, it would unofficially be known as "Task Force Faith." Lieutenant Colonel Faith completed his withdrawal south and joined with Lieutenant Colonel Reilly's 3rd Battalion at Pungnyuri Inlet by 1:00 pm.

Marine Captain Stamford's requests for airdrops of supplies were answered at approximately 3:00 pm when a large airdrop was completed in Faith's area by two C-119 flying boxcars. An hour after the airdrop, a Marine helicopter, requested by General Hodes at Hagaru, landed to evacuate the wounded. It left with Colonels Reilly and Embree who had been wounded on the night of the twenty-seventh. The enlisted men were furious that the officers were being evacuated first. Despite so many wounded, only four men were flown out that day. Morale was as bad as were the chances of a successful breakout.

Meanwhile just four miles away, Captain Drake's tank company made a second and final attempt to join the 31st Regiment. Even with air support, his tanks failed to take the critically important Hill 1221. Drake's tank company would remain in the area one more day and then under new orders would withdraw to Hagaru. But no one in Task Force Faith knew anything of this due to horribly lacking communications.

That night, Lieutenant Colonel Faith and his soldiers waited, and then around midnight, the dreaded bugles and cymbals announced another Chinese attack. Fighting continued all through the night but not as fiercely as the evening before. The bitter, brutal cold was as great an enemy as the Chinese.

HAGARU

At 6:00 am after a night of little or no sleep, General Smith sat on the edge of his cot. After hearing the reports of fighting throughout the night, Smith was very concerned about holding Hagaru. The thought kept running through his head: I will not be the first Marine general to lose a division!

After a meager breakfast, Smith put on his heavy parka and helmet and walked outside. The frigid air stung his face. Waiting was a jeep and a driver.

"Good morning, General," said the driver.

"Morning, Corporal," said Smith. "Take me to Colonel Ridge's CP."

Smith climbed into the jeep for the short ride to Ridge's command post which was a large tent. Even with the heat of two diesel fueled stoves, it was still too cold for Smith to take off his parka or gloves inside the tent. Smith walked to the back of the tent where Ridge was talking on the radio. Ridge acknowledged Smith's presence and held up his finger indicating he would finish his conversation quickly. While waiting, Smith looked around and saw the damage done by Chinese guns which had killed the battalion logistics officer and wounded the supply sergeant in this very command post the night before. Ridge signed off and turned his full attention to Smith.

"General, it was a rough night. As you can see, we took casualties all over the perimeter. There were some breakthroughs in the southern sectors at H and I Companies but they drove the Chinese back. Captain Corley of H Company was wounded but he is still in command. His executive officer was also wounded and two of his platoon leaders were killed. Including these officers, they had a total of sixteen killed and thirty-nine wounded. Fisher's I Company also took a beating. His executive officer was killed and four of his platoon leaders were wounded. Total casualties for Fisher were two dead and twenty-two wounded. Unfortunately, we lost a lot of officers and senior enlisted men."

"Colonel, what is the status on East Hill?" asked Smith.

"The fighting was fierce and we were pushed off the top but we still hold part of the hill. Sir, I know the casualties were very heavy on East Hill, but I don't have the numbers yet. I was just on the radio with PFC Podolak. When the Chinese overran the Army units on the hill last night, Marine Captain Shelnutt, who was commanding that unit, was killed. Podolak was his radio operator and this young Marine hid in a hole behind the Chinese lines. He continued to call in all night and gave me reports on enemy activity. He also helped us direct artillery and mortar fire. Incredibly, he's still stuck up there on that hill, all by himself."

"Colonel, we have to rescue that Marine!" exclaimed Smith.

"Yes sir. I agree and I will!"

"Colonel Ridge, with the few troops you had available, you did a fine job last night."

"Thank you sir, but with all the casualties we've taken, we're going to need replacements to hold this place."

"I've already ordered Colonel Puller to open up the road to Hagaru and get some troops up here. He is sending about 900 men. I expect them to arrive this afternoon. I also told Colonel Litzenberg to clear the road between here and Yudam-ni. I'm on my way to check on the casualties. Keep me posted."

Smith hoped his assurances of coming replacements would not prove to be cockeyed optimism. He had heard from Puller and Litzenberg about the forces they were facing.

He left Ridge and drove over to C Medical Company which was housed in an old school building 300 yards from the front line, south of the strip behind I Company. He walked into the building and was greeted by the chaotic sight of the doctors and corpsmen feverishly working on the wounded Marines. Smith just stood there; he didn't want to interfere. Lieutenant Commander Streit, the CO of C Medical, had just finished sewing up a young Marines' arm and saw Smith standing there. He looked over his shoulder and said, "Good morning, General."

Smith could tell that this facility was operating far in excess of its capacity. Casualties were everywhere; on stretchers, on the floor and on tables. Any flat surface was filled. Smith made his way through the chaos. Exhaustion showed on the faces of all the medical personnel who had worked throughout the night. No sleep, no coffee breaks, these men were dying. Their job was to save them. This was no eight-to-five job for anyone here.

"Good morning, Doctor," greeted Smith. "How many casualties do we have?"

"General, between C Med and E Med, we took in about 300 last night but they're still coming in," replied the weary doctor. "Some of these grenade and mortar wounds are so severe, we had no choice but to amputate. Under better circumstances, we probably could have saved many of these limbs, if only we had more doctors and more time. There is no way that medical schools could have prepared these young surgeons for these gruesome combat wounds. But somehow they're dealing with it."

"Is there anything I can do?"

"Sir, we are doing the best we can but it's not safe here. Last night we operated under strict blackout conditions. Bullets were coming through this place all night but fortunately, no one was hit. General, the Chinese shot up about half of our vehicles. Transporting the wounded is now even more difficult. In an hour, we're going to start moving these Marines to E Med where it's safer. Sir, we could sure use some more doctors and corpsmen."

"I'll see if we can helicopter some medical people up here from Hungnam. Doctor, I know you've got a lot of work to do and I won't keep you from it. I'm going to check on my men."

Smith walked through the ward and talked with some of the men. He had seen plenty of wounded on Peleliu and Okinawa. He understood that war meant death, but couldn't help but believe that the carnage he was seeing now was because of this reckless, senseless race to the Yalu.

Smith then drove over to E Med, the other medical facility, which was housed in a one story building in the center of town. It was in a much safer location than C Med. When Smith walked into the medical building, Marine engineers were in the process of constructing rough wooden racks on which stretchers could be stacked three high. This building would soon be overflowing with casualties.

The general removed his helmet, but not his parka: it was much warmer in here than outside. But in reality, the temperature inside the medical facility struggled to reach fifty degrees. Because of the constant bitter cold, the diesel stoves had a difficult time in keeping the makeshift hospital heated. The constant ins and outs of the engineers and medical people caused repeated blasts of Arctic air.

After talking to the doctors and visiting his men, Smith returned to his command post, still furious over the wasted lives he had just seen. The airfield had to open soon to help evacuate the wounded.

At noon, Army Brigadier General Hodes entered Smith's smoke-filled command tent.

"General Smith," he started, "I got more news on what's happening over on the east side of the reservoir. They've suffered over 400 casualties and they're in bad shape. In my opinion, General, I don't think they are going to make it unless they get some help. I know you've got your hands full, but is there any support you can provide?"

For Smith, the news was all bad.

"General Hodes, I know just what you are feeling," replied Smith. "We also lost a lot of good men last night. We need help and we need it right here. If I had all my Marines at Hagaru, I'd give you all the help you need. But as you know, I'm stretched pretty thin and there is nothing I can do. We have had several Chinese attacks and the enemy has even broken through our thinly manned lines. Fortunately, we have been able to regroup and hold, but we don't have a man to spare."

Hodes hung his head, understanding what Smith was saying but unable to accept what this meant for his men.

Both the Chinese and Americans knew that the key to holding Hagaru would be East Hill. The hill overlooked the packed Marine base and if the Chinese controlled the top, they could fire directly into its crowded perimeter. Within firing range were the ammunition dump, two medical centers, the engineers working on the airstrip and the division command post. These were all tempting targets. The Marines knew they had to neutralize the Chinese on the heights.

The Savior

Earlier that morning, Colonel Ridge ordered his executive officer, Major Reginald Myers to round up all the men he could find and take back East Hill. Not only the strategic position of the site, but the knowledge of brave PFC Podolak stuck in a frigid hole behind enemy lines motivated these men. Myers cobbled together a group of 300 Marines and Army support troops to attack the hill.

At 9:15 am, Major Myers made radio contact with Podolak.

"What's going on up there?" asked Myers.

"Sir," whispered Podolak, "there's Chinese all over the damn place and I sure would appreciate it if the Major would hurry up."

"We're on our way, Marine," responded Myers.

Fifteen minutes later, Corsair fighter bombers roared overhead and blasted East Hill with rockets, napalm and machine gun fire. Podolak burrowed as deep as he could in his hiding place while the ground shook from the armaments raining down on it.

Simultaneously, Myers started his ground attack up the hill. As the Marines fought and clawed their way up the steep, icy, 600-foot hill, they passed PFC Podolak's hiding place. Overjoyed at the sight of his fellow Marines, he jumped up and joined them in the attack.

Shortly thereafter, Podolak took a bullet in the back. But the bullet had gone through his radio pack thus saving his life. Although wounded, he pressed on with Myers' men as long as he could. Finally, he was taken down the hill to an aid station.

Major Myers had started out with over 300 men but less than 70 had made it near the top. The Marines reached the military crest, but the topographical crest was still held by the Chinese. When darkness settled over the hill, Myers set up a defensive position. He was barely holding on.

KOTO-RI

At 9:45 am, Smith had called Puller and told him it was urgent that Task Force Drysdale fight their way through to Hagaru. Puller replied that the badly needed replacements were just leaving. Nine hundred and ninety-two men left the perimeter at Koto-ri including the Royal Marines, Marine Captain Carl Sitter's G Company, and an Army rifle company on its way to join the 31st Regiment on the east side of the reservoir. The task force contained 141 vehicles. The Royal Marines and Sitter's G Company would attack the hills in leap-frogging style, while the Army Company would follow on the road. Air support would be provided by General Harris' 1st Marine Air Wing. Corsairs dropped napalm and the planes peppered the hills with machine guns and rockets. Three hours later the task force had only moved two miles because of heavy fighting. They still had nine more miles to go. Elements of an entire division of the Chinese army were positioned in bunkers on the high ground from Koto-ri to Hagaru. Drysdale was outnumbered nearly ten to one.

At 11:30 am, Puller called Drysdale to advise him that he was sending a company of tanks to assist him with his attack up the road. Puller told him that the tanks would arrive about 1:00 pm. Drysdale made the decision to halt the advance and hold the ground taken until the tanks arrived. Then he would push on to Hagaru. When the tanks arrived, they spearheaded the attack.

By 4:15 pm, they had only advanced two more miles when they came under heavy fire from automatic weapons and mortars. The tank commander reported that farther movement up the road was not advisable. Because the tanks blocked the road, all vehicles behind were stopped. Drysdale called Puller, explained the situation and asked what he should do. Puller called Smith and the general told Puller that Drysdale must fight his way through at all cost. Smith hated to give the order knowing that good men would die, but as a commander he also knew that many more good men would die if these reinforcements could not get through to help hold Hagaru.

Drysdale flinched when he got the order from Smith on a tank radio, but he, too, was a proud Marine. "Let's give it a bloody go," he told his men.

The task force resumed its attack. When dark came, the task force lost their air cover and the Chinese attacked in force, splitting up the long line of trucks into small, separated segments.

Well after dark, Marine Captain Carl Sitter broke through and led his G Company into the Marine defensive perimeter at Hagaru and reported in to Lieutenant Colonel Ridge.

"Sir, we had a tough time fighting our way up here. Lieutenant Colonel Drysdale was wounded and I assumed command of the task force."

Ridge interrupted the captain, "What condition is Colonel Drysdale in?"

"Sir, he took a bullet to the arm but I think he'll be okay. We are the lead element and the British Commandos will be following shortly. We are pretty strung out, the fighting was really heavy and because of the mountains, the communication in the task force was poor. It felt like all of China was out there shooting at us."

Colonel Ridge said, "Carl, I'm putting your company and the 41st Commandos in reserve for the night. Get your men some hot chow and some sleep. You'll be going back into action tomorrow morning."

Unbeknownst to Sitter, the Marines and soldiers in the middle of the column were trapped and accounted for most of the missing in action. At 4:00 am, cut off and out of ammunition, they had no choice other than to surrender. Those at the rear of the column fought their way back to Koto-ri by dawn.

It had been a terrible battle for the 922 men who had started out. Added to the tiny forces at Hagaru were a tank company of 100 men and 300 seasoned Marines and British Commandos. A total of 300 Marines made it back to Koto-ri. The rest were either killed or missing and presumed captured. Less than half of the badly needed replacements that Smith had hoped for actually made it through. Many of those that did were wounded or frostbitten from their day-long battle, thus adding to the burden on the medical staff.

Of the 141 vehicles in the convoy, seventy-five were destroyed and one tank was lost. Drysdale would refer to this area of battle as "Hell Fire Valley."

Shortly after the sun went down, the Chinese attacked Puller's fortified perimeter at Koto-ri. They came charging down from the hills, in seemingly endless numbers, on the northern end of the perimeter and assaulted Captain Jack Smith's E Company. Seventeen enemy soldiers got inside the perimeter but were killed. When the Chinese pulled back an hour later, they left 175 dead in the snow and an estimated 200 wounded. The Marines lost six killed in action and eighteen wounded. Puller would hold at Koto-ri.

HAGARU

Late in the evening, Smith was in the command tent when Almond radioed in new orders. Smith was told to stop the attack to the northwest. The drive to Yalu was finally, officially over. Smith was disgusted at the irony of these orders. In addition, he was given command of all Army units in the Chosin Reservoir area including all troops on the eastern side of the reservoir. He also was told to deploy one regiment from Yudam-ni to Hagaru and then gain contact and assist the Army on the east side and concentrate all forces at Hagaru. Clearly, easier said than done. How many lives have already been wasted because of the ego and ambition of one man?

After thinking about this for a minute, Smith told Colonel Bowser to follow him to his quarters. They walked out into the bitter cold; their breath blowing out a white cloud of vapor. It was a very short walk to Smith's house. When they walked in the door Smith motioned Bowser to sit down.

Bowser spoke first. "General Almond dumped this big mess right in your lap."

"Yes, he did Al, but the reality is, we have to deal with it. Here we are in the mountains of North Korea and we're fighting the Chinese over this worthless piece of ground and this country doesn't belong to either of us."

Smith lit his pipe and continued, "Did you ever wonder why we're stuck up here in these miserable mountains? It's because the people in power in Washington and that includes the Joint Chiefs of Staff, didn't have the courage to say no to 'the General' in Tokyo. He wants to capture real estate for publicity without understanding how worthless the real estate is or how many lives of American servicemen it would cost.

"Alright, let's take a look at Almond's new order. Litzenberg and Murray have already cancelled the attack north yesterday; so that's done. I don't know how we're going to get a regiment from Yudam-ni to help those soldiers on the east side. This is just unrealistic. Before we can do anything Litzenberg and Murray have to fight their way down here and then and only then can we help the Army on the east side. Al, have Litzenberg and Murray draw up a joint withdrawal order."

Smith continued, "Al, if General Craig was here I'd send him up to Yudam-ni to take over, but we obviously don't have that option."

"Sir, Litzenberg and Murray are doing a good job up there."

"I know they are working well together, but it would have been better to have General Craig in command. I guess I wouldn't have let him go if I'd known we would be surrounded," replied Smith.

"Al, I owe you an explanation." Bowser tried to say something but Smith held up his hand to stop.

"Do you know anything about General Craig?"

"I know he served in Nicaragua and in the Pacific in World War II and was awarded the Navy Cross and of course he led the 1st Marine Brigade in the Pusan Perimeter at the start of the Korean War. I'd have to say sir that he's an outstanding Marine."

Smith took a pouch of Sir Walter Raleigh tobacco out of his pocket and started filling his pipe and then he lit it. "Let's go back a few years. General Craig met his wife in Shanghai. She was a British citizen. Fast forward to World War II and Colonel Craig is now the commanding officer of the 9th Marines. As he is about to leave the United States, his wife contracted a bad case of TB and she tells him to go; it's his duty and he has trained for this moment. Shortly after he arrives in the Pacific, he receives a letter from her doctor saying that his wife has taken a turn for the worse and that he should fly home. He showed the letter to the division commander, General Barrett, who said there was no reason why he couldn't go home as the division would not be in combat for a few months. However, Major General Vogel who commanded the Amphibious Corps, Pacific Fleet, to which the division was attached, refused to approve the leave. General Craig's wife passed away shortly thereafter. He never got to see her. Al, I couldn't let that happen again."

Smith brought the pipe up to his mouth and took a couple of pulls on it and then he said, "General Craig found out later that General Vogel had sent a plane to the United States a few days later to pick up a load of liquor and a bar. Also, several staff officers were on that plane and they got to visit their families. Craig told me he lost all respect for Vogel. General Vogel was relieved later because he could not carry out his command. After that experience I thought it was important that Craig be able to get home before his father died. I thought you should know some of the background to my decision."

Colonel Bowser stood up, put on his parka and said, "Sir, I would have done the same thing for General Craig. It was the right thing to do. We'll get through this. We have good men out there. Good night sir." Bowser walked out the door.

The Chinese were strangely quiet that night at Hagaru and Yudam-ni.

TOKTONG PASS

Captain Barber found it necessary to shrink the size of his defensive perimeter due to the number of casualties that he suffered. Care of the wounded was difficult in the freezing weather. Helicopters tried to evacuate the wounded but were being hit by small arms fire and could not land.

Airdrops of supplies continued through the day. One drop was 600 yards off the mark and Barber's men had to lay down covering fire in order to retrieve the wayward supplies.

At 2:00 in the morning, the Chinese attacked again. But this time, they attacked from the south, crossing the road and charging up the hill. Barber's machine gunners and riflemen opened up and cut the Chinese to pieces. The battle only lasted ten minutes but the losses to the Chinese were staggering. Over 400 Chinese soldiers lost their lives while Captain Barber lost only one of his Marines.

The next morning looking down the frozen hill now littered by bodies, Barber and his men saw a grotesque scene of icicles of blood. Red rivulets ran from each body for a few inches before the cold froze the blood. The red lines stood in stark contrast to the otherwise white landscape. Barber again had his men collect many of the frozen Chinese bodies to stack around their fighting holes as protective barricades. It was gruesome work but given the shallow fox holes, the men knew that any protection was better than no protection.

This is a helluva way to celebrate my birthday, Barber thought to himself. He had just turned thirty-one years old.

12

THE SURVIVAL OF X CORPS

November 30

HAGARU

It was another busy day for Smith. Early in the morning he met with Bowser in the command tent. General Smith lit his first pipe of the morning and said, "Al, I'm really concerned with the situation on East Hill. We have to control that ground."

"Sir, I just spoke with Lieutenant Colonel Ridge. He is sending Captain Sitter's G Company to replace Major Myers' Marines. Myers' men were pretty well shot up yesterday. G Company isn't in great shape after the trek yesterday from Koto-ri but they will help. Colonel Ridge will hold the British Commandos in reserve."

At 8:00 am, Captain Sitter's G Company assembled at the bottom of East Hill waiting to attack. His men had gotten little sleep from the night before and were still cold and exhausted after fighting their way from Koto-ri to Hagaru.

As they started up the hill, their lungs and throats burned from the frigid air. Their eyeballs felt like they were being sand blasted and the hair in their noses froze. Progress was painfully slow because the hill had been pounded into a sheet of ice by those who had proceeded them. The Chinese took full advantage of their position on top of the hill and poured deadly automatic weapons fire on the Marines. The daylong attack bogged down.

The Savior

Back in the command post, Smith was greatly concerned about the deteriorating situation in his area of operations. The general sat quiet for a moment, deep in thought. Then he said, "Al, I want you to get someone to fly up to Yudam-ni. I want some firsthand information on what's happening to Litzenberg and Murray. Our communications are poor at best."

"General, I'll fly up there right away."

"Absolutely not," replied Smith. "Send one of your men. I need you here."

"All right, sir. I'll have my assistant operations officer, Lieutenant Colonel Winecoff, helicopter up there immediately."

A few minutes later, General Hodes came into the command tent and greeted the Marine general. Smith shook his hand and after offering the Army general a hot cup of coffee and a chair, Smith started, "General Hodes, as you are aware, General Almond has given me command of all Army units in the Chosin Reservoir area. I'd like you to draw up a dispatch in my name to the forces on the east side. I was told today that Colonel MacLean is either dead or missing and that Lieutenant Colonel Faith is now in charge. Is that correct?"

"Yes, Faith is in command," said Hodes.

"General Smith, could I use one of your clerks?" asked Hodes.

"Certainly," replied Smith as he motioned for one of his clerks. "Corporal, General Hodes is going to dictate an order and I want you to type it up."

General Hodes began, "Lieutenant Colonel Faith and his command are now attached to the 1st Marine Division. Every effort should be made to move south to Hagaru. You should do nothing which would jeopardize the safety of the wounded. You are authorized to destroy such equipment as would impede your movement. In addition, in view of the critical requirements for troops to hold Hagaru, no actual troop assistance can be furnished. But, weather permitting, unlimited air support will be available to the task force to assist in its movement southward."

Hodes looked at Smith and asked, "Anything else General?"

Smith replied, "Nope, that's it. We both wish it could be a lot more for those men but it's all we can do."

Fifteen minutes later, General Barr, the CO of the Army's 7th Infantry Division, flew into Hagaru and was immediately jeeped to Smith's CP. Barr and Smith had known each other since the planning meetings for the Inchon landing. Both were infantry officers and they had taken a respectful liking to each other. Barr had spent a couple of years with the Chinese, as a military advisor to Chiang Kai-shek; so he had a good sense of the enemy here in Korea.

"Good morning, O.P.," greeted General Barr.

"Welcome to Hagaru, Dave," answered Smith.

"O.P., I'd like to borrow a helicopter from you to visit Lieutenant Colonel Faith to advise him of the change in command."

"I think that would be a good idea. Better not wait too long. The snow is due to start accumulating soon," said Smith and handed Army General Barr the typed dispatch which Hodes had signed.

Smith, Barr and Hodes conferred for a few more minutes before Barr left to fly to Sinhung-ni for the visit with Faith.

After his meeting with Faith and finding out just how dire their situation on the east side was, General Barr returned to Hagaru to update General Smith at his command post.

Barr was greatly concerned about his isolated regiment.

Shaking the snow off his jacket, he said, "My God, O.P., I had no idea. They've got 500 casualties and who knows how many more when they start to withdraw. O.P. is there any way we can help them out?"

Smith looked over to Colonel Bowser and said, "Colonel, bring that map of the Chosin Reservoir area over here."

Bowser laid it on the table and Smith and Barr stood to look at it.

"Dave, I've got two regiments of Marines up at Yudam-ni who are encircled and fighting for their very survival and you've got a regiment on the east side in a similar situation. The only way out for either of these outfits is to fight their way to Hagaru."

Smith pointed to Toktong Pass, "I've got a company of Marines up there. They are surrounded, vastly outnumbered, and they are just barely holding on. They can't withdraw to Hagaru or Yudamni because they have too many casualties. Dave, I can't help them either. We have to hold Hagaru! As I told General Hodes, it's not that I won't help you; I just can't help you. Hagaru is the linchpin of this entire withdrawal. All is lost if Hagaru falls. And if I know anything about these Chinese, they have Hagaru right in the middle of their sights. I pray that this small group of men can defend Hagaru until the Marines and soldiers fight their way to us."

At 12:30 that afternoon, Almond flew into Hagaru for a conference with the commanders of the Chosin Reservoir area. Generals Smith, Barr and Hodes were present. The site of the conference was at the Hagaru airstrip in two connected pyramid tents. Diesel stoves in the tents gave only minimal warmth.

Passing two Marines guarding the tents, Almond stopped. The Marines hadn't bathed in weeks. They had only slept a few hours in the last few days. They were beyond cold.

General Almond, in his clean parka and pressed winter trousers must have heard some talk that he wasn't known for his rapport with the troops. He decided to try some small talk with the two Marines.

"Mighty cold out here. My dental plate almost froze in the water glass by my bed last night," he said.

"That's a fuckin shame, sir," muttered one Marine who hadn't been in a bed in months.

Almond said nothing but gave the Marine a hard look as he entered the tent.

Almond started the conversation, "Good afternoon gentlemen. First off, General Smith, I want the 5th and 7th Marine regiments withdrawn from Yudam-ni immediately."

On the twenty-ninth, Smith on his own initiative, had already ordered Litzenberg and Murray to start planning a joint withdrawal order.

"Yes sir," responded Smith. "I have a staff officer up there with them now planning a withdrawal."

"Next," said Almond, "I want Generals Smith and Barr to submit a plan to extricate Task Force Faith from the east side of the reservoir."

Almond then looked at Barr and said, "If Faith doesn't execute these orders, I want him relieved."

Smith couldn't believe what he was hearing.

General Barr quickly replied, "I strongly object to that General Almond. Lieutenant Colonel Faith is an outstanding officer and I have a lot of confidence in him. I was just with him and as I have reported, he is facing what looks like the whole Chinese army, with hundreds of wounded and no safe exit route. He is doing the absolute best that can be done under these circumstances."

Almond made no response. Smith wondered if anything Barr said was penetrating the X Corps commander's brain.

General Almond continued, "X Corps will abandon the Chosin Reservoir area when all troops have fought their way to Hagaru. Once there, they will withdraw south to the coast."

It was apparent to Smith that Almond was an entirely different man from the overly confident one he had been when he visited Hagaru just two days earlier. But Smith was not convinced that he was any more capable as a leader. He didn't seem to have a firm grasp on what the "fight" to Hagaru required.

But at least Smith thought that Almond finally realized that the very survival of X Corps was at stake.

Almond went on, "General Smith, speed is of the utmost importance. You have my promise that the Marines will be resupplied by air. You have my approval, if you choose, to burn or destroy as much of your equipment as you think necessary in order to hasten your withdrawal."

Smith just didn't have much patience with Almond's limited understanding of the tactical situation and curtly replied, "General, my movement will be governed by my ability to evacuate the wounded and to clear the road south. It's sixty-four miles to Hungnam and I'm going to have to fight every mile on my way back. I can't afford to discard my equipment."

Ignoring the verbal jab, Almond then told Smith to submit a plan, with timetable, for the extraction of Task Force Faith. No further discussion was necessary or forthcoming.

Almond left and flew back to Hungnam.

Smith and Barr had already talked in great detail about Faith's predicament and they both agreed that under the circumstances a timetable made no sense until the Marines at Yudam-ni fought their way back to Hagaru.

"It's just bullshit paperwork," said Barr. "We might as well draw up a timetable to invade China."

General Barr left and flew back to Hungnam. He would try again to break the news to Almond that there was no workable plan or timetable to help Faith.

Smith left the airstrip and was driven back to the command tent where he met Colonel Bowser. Smith started to vent.

"Almond just told me to destroy or burn all our equipment. Al, he just doesn't get it. In my opinion, he has lost all his effectiveness as a Corps commander. Let's face it, we're pretty much on our own now and anything X Corps has to say is certainly suspect. Al, we're going to have to fight our way out of this mess without the support of X Corps. The best we can hope for is that they don't get in our way."

Just then an aide to Smith said, "Sir, Lieutenant Colonel Winecoff has just returned from Yudam-ni."

Smith said, "Send Winecoff in."

A few minutes later, Winecoff entered the command tent and Smith asked, "What's going on up there?"

Handing General Smith the joint withdrawal order drawn up by Litzenberg and Murray, Winecoff started, "Sir, the situation up there is far worse than I could have imagined. They have at least 500 casualties; there is also a growing number of frostbite cases and this is only going to get worse. Yudam-ni might as well be in the Arctic Circle, it is so cold up there. Their weapons are freezing. They are oiling their rifles with hair tonic.

"They suffered most of their casualties on the night of the twenty-seventh. Although not as severe, they were hit again on the twenty-eighth and twenty-ninth. The fighting has been intense and the number of Chinese seems limitless. These men are badly shaken up and they feel like they are isolated up there. Helicopters are flying out the most seriously wounded Marines. But only being able to fly two men out at a time is just a drop in the bucket. The casualty list keeps growing. The Chinese and the extreme cold are chipping away at them."

"Let me see that joint withdrawal order, Colonel," requested General Smith.

Smith quickly read the order. The first part of the plan called for a readjustment of the perimeter at Yudam-ni. The plan called for a gradual withdrawal from the north and west of Yudam-ni by the 5th Marines so they could relieve the 7th Marines. The 7th Marines in turn could move south of the village. The second part of the plan called for the relief of Captain Barber's F Company who was barely holding on at Toktong Pass.

Lieutenant Colonel Ray Davis, CO of the 1st Battalion 7th Marines was given this job.

The plan called for him to hike through the icy mountains under the cover of darkness and reach and rescue the battered Marines at Toktong Pass. Meanwhile, the 5th and 7th Marines would leave Yudam-ni and fight their way to Toktong Pass. Once there, they would hook up with Colonel Davis and Captain Barber and fight their way south to Hagaru. All wounded would be transported south by truck. All other Marines not sick or wounded would walk.

The breakout would start first thing in the morning.

Smith glanced at his watch. Almost five o'clock and it's already dark. He left the command tent and walked briskly to his quarters for dinner. He couldn't help but notice the constant drone of the engines of the jeeps and trucks. Night and day these vehicles had to run to keep them from freezing. That used a lot of fuel, but Smith had stockpiled sufficient supplies at Hagaru and the airdrops were continuing.

After eating, Smith returned to his command tent and later he and Bowser again reviewed the withdrawal plan for Yudam-ni. He knew that giving an order and actually making the plan happen were two different things.

Around midnight the second major attack on Hagaru began. Again the Chinese tried to break through the southwestern perimeter line held by Lieutenant Fisher's I Company. The Chinese had made a very big mistake. Fisher's company was well armed and well prepared. He had built up a creative network of barbed wire, trip flares and booby traps. His men were protected by sand-bagged fighting holes and weapon emplacements.

When the attack began, each Chinese assault wave was mowed down by the unbelievable fire power that a Marine rifle company, reinforced by artillery, tanks, mortars and heavy machine guns, could bring to bear. They killed over 500 Chinese and left many more wounded lying in the snow. The frozen Chinese bodies were strewn across the landscape some sitting upright caught in the snow covered barbwire. Fisher had two men killed and ten wounded.

Over on East Hill, Captain Sitter was barely holding on. When it became dark, he had no choice but to set up a defensive line on the ground previously held by Major Myers, who by this time, had withdrawn his badly depleted force.

Although wounded by a grenade in his face, arms and chest, Captain Sitter remained in command and personally visited each machine gun position and fox hole of G Company and its reinforcements. Because of his bold leadership, Sitter inspired every man on East Hill to hold on.

Later, the night became bright as day when an enemy artillery shell blew up fifty gallons of gasoline in a supply dump. This lit up the battle so brilliantly that General Smith could see it from the doorway of his command tent almost 1,200 yards away. He could also see the red lines of the tracers from the rifles and machine guns. He heard the booming sounds of heavy artillery and the constant chatter of the machine guns. The smell of burnt gunpowder filled the air. Smith knew that although he was badly outnumbered, his Marines had to hold Hagaru.

Defeat was not an option!

EAST SIDE OF CHOSIN

Chinese attacks on Faith's perimeter continued all afternoon and into the night. Word was passed among the perimeter units that "if we can hold one more night, we'll have it made."

Just shortly after midnight the intensity of the Chinese attacks reached heretofore unknown levels. Holding on for one more night would come at great cost for both sides.

The Chinese were determined to overrun the defensive perimeter; the soldiers were desperately determined to prevent it. Fierce fighting continued all night with many lives taken and still many more wounded and dying. Burning vehicles cast eerie shadows. The dead and the wounded were hopelessly intermingled. The estimated count of 500 wounded to be evacuated rose very sharply as the number of fighting men dwindled.

TOKTONG PASS

Thick clouds moved in over Toktong Pass as Captain Barber's Marines stared out into total darkness. Shortly before midnight, four inches of fresh snow began to accumulate.

The Savior

Taking advantage of the reduced visibility the Chinese, on the high hill to the north, moved four machine guns into position and fired at the Marines.

Barber radioed Hagaru to request artillery fire. He would illuminate the target with his 81mm mortars. In a rare display of totally accurate artillery fire, the four rounds of the first salvo, coming seven miles away from Hagaru, landed directly on the machine guns. Pieces of bodies and gun barrels flew through the falling snow. This stopped the Chinese dead in their tracks. The shocked Chinese troops turned and ran. The rest of the night remained relatively quiet.

Barber's desperate and depleted company of Marines would hold for yet another night!

13

THE TURNING POINT

December 1

HAGARU

At nine o'clock in the morning, Lieutenant Colonel Ridge reported into General Smith. Smith could tell right away that the strain of defending Hagaru coupled with almost no sleep was wearing Ridge down.

"Sir, if the Chinese attack again in force like they did on the twenty-eighth and last night," started Ridge, "I'm not sure we can hold them. We just don't have the forces available. I don't see how we can hold both the airstrip and East Hill. With the casualties increasing both from combat wounds and frostbite, our resources are dwindling."

Smith leaned in toward Ridge and said, "Colonel, I understand what you are saying, but we don't have a choice. We have to hold both these points and we will hold them with what we have until Murray and Litzenberg fight their way down here to join us. If we don't hold, they will have no place to go."

"Yes sir," replied Ridge.

"Colonel Ridge, hang on; we'll get through this."

Ridge left and went out to buck up his men. He was dead tired and too cold to even remember what it was like to be warm. But he was a Marine and his general was right there with him.

Not long after Ridge left, Navy Captain Eugene Hering, the division surgeon, stopped in to see Smith with more bad news.

Hering was responsible for the planning, organization, and execution of all medical operations of the division. Fortunately, he was not new to combat. In January 1944, Hering became the Division Surgeon of the 2nd Marine Division and served in combat at Saipan, Tinian, and Okinawa. He was awarded his first Legion of Merit for his outstanding performance. When the Korean War started he volunteered to command the Medical Section of the 1st Provisional Marine Brigade. For his service in the Pusan Perimeter in August and early September he received his second Legion of Merit. For his extraordinary performance at Inchon came a third Legion of Merit. When the 1st Marine Division arrived in Korea, General Smith appointed Hering the Division Surgeon.

"Good morning, General," greeted Hering. "Do you have a moment? I need to talk to you about the casualties."

"Good morning, Doctor," replied Smith. "Certainly, please sit down. How many casualties do you have?"

The general sensed he was about to get an earful of unpleasant news.

"Sir, we have over 600 here at Hagaru. But here's what scares the hell out of me. When Task Force Faith breaks out, they are going to bring possibly 500 additional casualties or maybe more. Then, when the 5th and 7th Marines fight their way down here, they could bring possibly 500 more. General Smith, that's 1,600 casualties and I think that's a conservative estimate. There's no way we can handle that many with the resources we have."

"Doctor, your people are doing an exceptional job of taking care of the wounded," praised Smith.

Dr. Hering responded, "General, it's those Navy Corpsmen up in the hills who are saving lives. When a Marine gets hit by shrapnel or a bullet, the corpsman is the first one to respond. They don't care about their own safety. They stop the bleeding; they ease the pain. You know the enemy takes special aim at the corpsmen. They are prized targets. Morale goes to hell fast when a corpsman is down."

"General, it's so cold out there that they have to carry the syrettes of morphine in their mouths or in their crotch to keep it from freezing. The blood plasma is freezing and can't be used. Some of the men are shot in the arm or leg and can walk in here unassisted; some wounds are minor enough that they can be patched up and returned to duty. But the Marines that are seriously wounded take four other Marines to carry them in here.

"The head and chest wounds are the toughest for us to deal with. In some cases the bitter cold has frozen the wound shut and saved the man's life."

"General," Hering continued, "the casualties in the medical companies are piling up and I have great concerns over our ability to provide adequate care. We can barely keep them warm much less provide proper treatment. And we're only dealing with serious injuries at this point. Just about everyone here has a bad cough and cold from living in this frozen wasteland. We aren't doing anything for these ailments and they're also taking a toll on your Marines."

"There is only one way to solve this problem," Smith said. "We've got to open the airstrip and fly these injured men out. And we will do our best to get them out of here."

Captain Hering left, relieved that his commanding officer understood the situation and was going to try and fix it.

Smith and Bowser talked for a while about the seriousness of the increasing casualty situation. Finally Smith told Bowser to have Lieutenant Colonel Partridge report to the command tent. Partridge could be a salty old dog to deal with and Smith didn't relish the conversation after having spent the day already getting bad news.

"Sir," Bowser replied, "Colonel Partridge just flew in from Hungnam and he's down at the strip meeting with his engineers. I'll get him up here right away."

The engineer walked through the tent door fifteen minutes later.

Smith asked, "John, how soon can we get that strip open?"

"Sir, we've been working around the clock for twelve straight days. I've got five large dozers and six tractors going non-stop to get this job done. As you know, my men have had to stop work a number of times to pick up their rifles and fight. The Chinese are doing all they can to stop us from building this airstrip. The runway is currently 2,900 feet long and fifty feet wide. It should be at least 4,000 feet."

"John, do you think a C-47 would risk landing on 2,900 feet?" asked Smith.

"If the strip was at a lower altitude, I think they could land on it; but we're about 3,500 feet up here. Sir, I'm am engineer, not a pilot. The air wing has to make that call."

"Our casualties are rapidly mounting and we've got to get them out of here," said Smith. "This two-at-a-time in helicopters just can't keep pace. We have hundreds of wounded who need to be evacuated."

"I understand, sir," responded Partridge. "I just ran into Dr. Hering and he let me know what he is up against."

"One other thing, sir," continued Partridge. "Since this strip is only fifty feet wide, it means only one plane can land and take off before another lands. There are no taxiways, but we will continue to lengthen the runway and create areas to park additional planes."

Partridge left and then Smith told Bowser to radio the Marine Air Wing and explain the casualty situation and see if they would attempt a trial landing with a C-47 transport plane. He knew it was a risky thing to ask of a pilot, but he was desperate.

Smith left the command tent and returned to his quarters. As the general was finishing a quick meal, Bowser threw open the door and said excitedly, "General, come quick! You've got to see this."

Smith was greeted with a beautiful sight. A squadron of C-119 "flying box cars" were air dropping much needed supplies.

Parachutes dotted the sky as rations, ammunition, medical supplies, and fuel drifted into the Hagaru perimeter. Marines all over the camp were cheering. This would be the first airdrop of supplies by the Air Force's Combat Cargo Command flying from Japan. They would from then on deliver almost 100 tons per day of needed supplies as weather permitted. Given how quickly the supplies were being used with all the Chinese attacks, the airdrops were critical. American air superiority was paying off.

General Smith turned to Colonel Bowser and said with a big smile, "Al, we have to give a lot of credit to the Air Force. They are really getting the job done."

Bowser replied, "Yes, sir. With the road closed, we would be in big trouble without them."

At 2:30 in the afternoon, Smith was driven to the airstrip. While he was there, an Air Force C-47 plane made a low pass over the partially completed field. Then he climbed back into the air, circled and made a second pass. The bulldozers and other heavy equipment at the site hurried to get off the strip.

A crowd started to gather as Marines left the relative warmth of their tents to see what the excitement was.

Smith tensed in anticipation.

"Please God," he prayed silently, "let him land safely."

The plane came in low and landed on the frozen, snow-covered strip. The big two-engine aircraft bounced and skidded its way over the uneven surface and came to a stop at the very end of the runway. But the plane made it.

Smith's driver, temporarily forgetting who his passenger was, blurted out, "Man, that son-of-a-bitch must have a set of brass balls."

Smith just grinned and mentally agreed with the assessment.

Smith watched as the first twenty-four casualties were quickly loaded aboard the plane. Half an hour later the big plane took off over the rough ground and slowly climbed into the air, barely clearing the hills, climbing higher to the wild cheering of the Marines.

Smith immediately recognized this as the turning point in the battle of Chosin Reservoir. Not only could the wounded be flown out but reinforcements and supplies could also be flown in. Colonel Partridge's engineers had accomplished the near impossible task of carving an airfield from the frozen ground in the mountains of North Korea. He may be a gruff fellow to deal with, but Smith wouldn't trade him for anything.

Smith returned to the command tent and told Colonel Bowser to spread the word that the strip was open.

"Hundreds of combat trained Marines remain in the rear areas such as Hamhung and Hungnam. They couldn't get up here when the road closed. Al, start flying these men in here. We have our reinforcements."

"I'll get right on that, General."

Over the next few days, 500 badly needed replacements would be flown into Hagaru to reinforce the battered infantry companies of the 1st Marine Division. Composite companies were created from reinforcements and remnants of hard hit companies.

That afternoon four of the C-47s landed at Hagaru. Three of those planes returned to the air loaded with sixty casualties in total. The fourth plane, heavily loaded with ammunition, collapsed its landing gear as it touched down on the bumpy strip. The plane was unloaded and pushed off the runway. It was getting dark and air operations were suspended for the night.

YUDAM-NI

At eight in the morning, the 5th and 7th Marines started their breakout to the south. It was only fourteen miles to Hagaru but it would take three days of intense bloody fighting to get there. At times it seemed like a thousand miles. The men were heavily bearded. They had no way to shave and found the facial hair gave a layer of insulation from the cold. They were physically exhausted, deprived of sleep, food, water, and sanitation, and were constantly battered by sub-zero temperatures. But they were Marines and they still had fight in them.

When the column started moving down the road, the Chinese started shooting at the drivers. When a driver was hit, it took a lot of courage for another man to take the wheel. But it happened over and over again as the convoy moved southward. The Marines were breaking out.

Litzenberg and Murray put their best battalions in the lead positions. The enemy would move up the slopes of the hills as the Marines advanced. The Marines would call in air strikes to take them out. The infantry would then take to the hills while the long truck column slowly moved down the road. The vehicles were loaded with the wounded as well as ammunition, rations and medical supplies. All additional space in the trucks would be needed for casualties incurred in the breakout as well as F Company casualties at Toktong Pass. That left no room to bring out the dead.

Litzenberg and Murray had made a difficult decision: the dead would have to be buried at Yudam-ni. Bulldozers carved a long burial pit out of the frozen ground about six feet deep. The eighty-five officers and men lost at Yudam-ni were wrapped in parachute cloth and put to rest in a mass grave just south of the village. The burial service was as hard on the men of Yudam-ni as the fighting had been. They were trained to leave no man behind and leaving their comrades in this frozen soil took its toll.

EAST SIDE OF CHOSIN

At nine in the morning, in a meeting with his officers, Army Lieutenant Colonel Faith agreed that today must be the day to fight their way from Sinhung-ni to Hagaru only nine miles away. At the same time, Marine Captain Stamford established radio contact with a lone Marine fighter bomber. The pilot advised Captain Stamford that if the weather held, he would guide in a flight of Corsairs about noon to provide desperately needed air support.

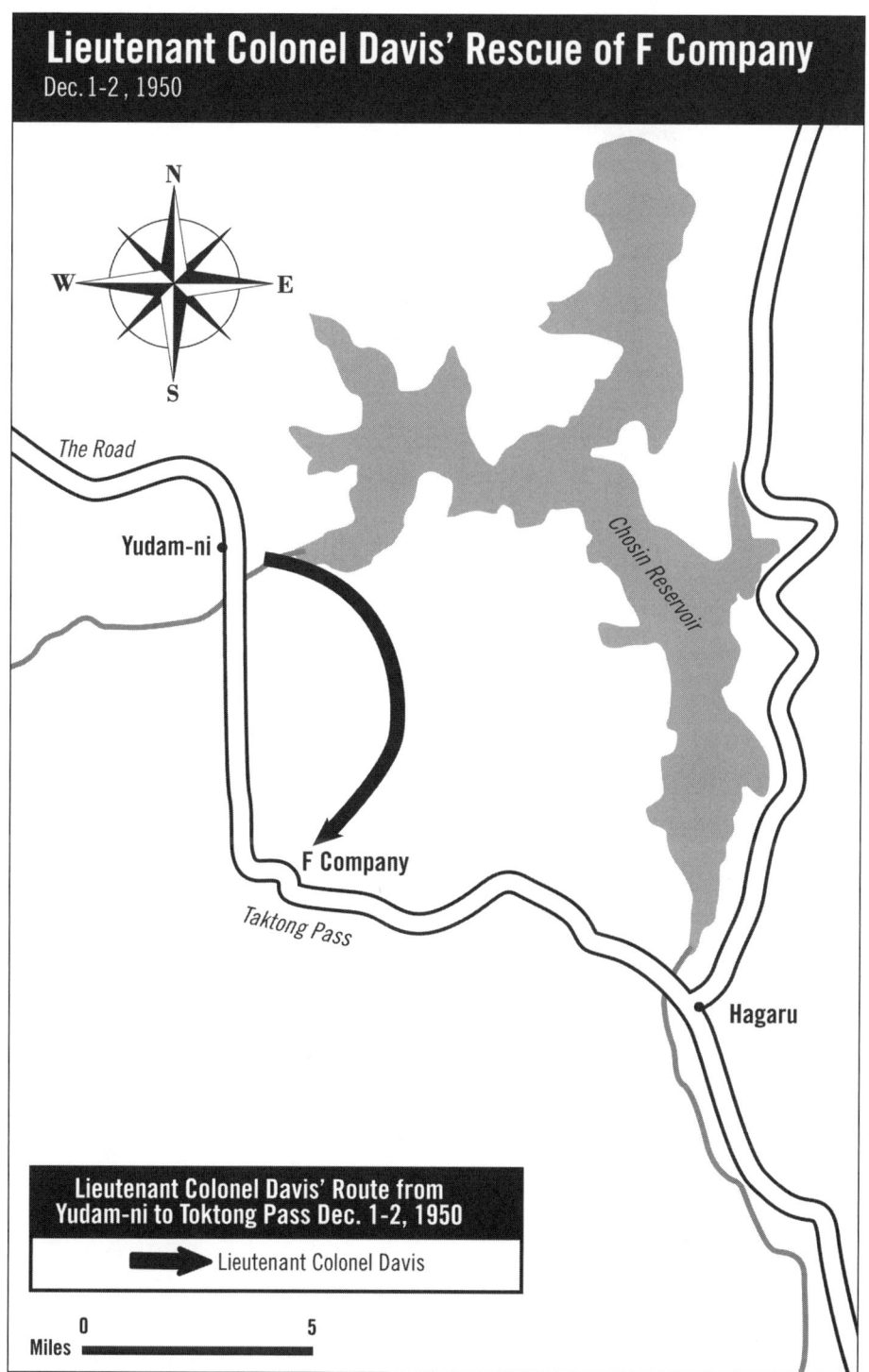

Lieutenant Colonel Davis' Rescue of F Company
Dec. 1-2, 1950

In preparation, all two-and-a-half ton trucks would be unloaded of their cargo and the wounded would be carried instead. All inoperative vehicles and jeeps were to be destroyed as well as unused supplies and clothing. Nothing useful was to be left for the Chinese. Infantry companies would lead the way and also provide protection for the column of the wounded.

At 1:00 pm, the Marine planes roared overhead. The withdrawal had begun, leaving behind piles of smoking supplies and equipment.

The Chinese had set up a series of roadblocks and blown bridges that took the Army units a long time to break through. The Chinese held the high ground on both sides of the road. The enemy attacks on the column were intense. Casualties started to climb.

As darkness fell and air support ended, Lieutenant Colonel Faith was killed as were many of the officers and senior enlisted men. They were only four-and-a-half miles from Hagaru but without leadership and air cover, the task force started to disintegrate. With Faith now dead, none of the Army officers stepped forward to take command. The Chinese continued to destroy the convoy truck by truck. As trucks became disabled they blocked the road. The drive for self-survival overcame the soldiers and it became every man for himself. All semblance of military discipline ceased to exist. The soldiers took off across the ice and the wounded in the truck column were abandoned. Army soldiers were leaving their wounded in overturned trucks to either freeze or be killed by the Chinese.

Realizing that all was lost, Marine Captain Stamford destroyed his radio to keep it from falling into enemy hands. During the night he was captured by the Chinese but escaped. He was careful to stay off the road and finally entered the Marine perimeter at Hagaru at about 2:30 am.

White phosphorous grenades could be seen and heard as they lit up the gruesome slaughter while the Chinese overran the unprotected trucks of the wounded. For some, death came quickly; for others, their torment continued until the cold, silent veil of death covered them and put them at peace. Sitting dead in his truck, Lieutenant Colonel Don Faith never lived long enough to see the end of Task Force Faith. Stragglers and escapees walked and crawled out onto the ice of Chosin Reservoir in hopes of finding safety with the Marines in Hagaru. The eastern side of the reservoir became eerily silent.

TOKTONG PASS

As part of the plan in the breakout from Yudam-ni, Lieutenant Colonel Ray Davis, the CO of 1st Battalion 7th Marines, would hike through the icy mountains at night with his battalion to rescue Captain Barber and his beleaguered F Company. Davis and Litzenberg had spent time together formulating the plan which both knew could easily turn into a suicide mission. There were real risks. The plan would take one of Litzenberg's top battalions away from the rest of his forces. Splitting his troops like this could compromise the whole breakout, but having the Chinese control Toktong Pass was a greater threat.

The trip would be eight miles through enemy held territory. The piercing wind would be both a blessing and a curse. It would keep down the noise of their trek, but would penetrate their bones. The march would put them so close to Chinese troops that Davis would later report to General Smith, "We could smell the garlic on their breath."

The Marines were overloaded: each man carried a sleeping bag, extra clothing, an extra bandoleer of ammunition and three days of rations. Most chose cans of fruit because the sugary syrup would quickly provide much needed energy. The only way to prevent the rations from freezing was to keep them close to their bodies. The dry rations of crackers and jam and chocolate helped provide some energy but was clearly not a diet to keep fighting men in good condition.

They took only two 81mm mortars and six heavy machine guns. Mortar and machine gun ammo were carried on stretchers. If casualties mounted and the ammo was used up, the stretchers would be put to another use.

Davis' battalion had been up since early morning; they were already exhausted and foot sore from a full day of fighting. Ahead of them lay miles of dark, frozen hills and an unknown number of Chinese soldiers.

When Colonel Davis' Marines started out at 9:00 pm, it was already twenty-four degrees below zero. Climbing up the steep, icy hills was absolutely brutal. For the first Marines in line, it wasn't so bad; but soon the path would get slick and icy and men would slip and fall. Going downhill was even worse, especially for the machine gunners and mortar men with their unbelievably heavy loads. Often times the snow would be knee deep and sometimes waist deep and it sapped the energy out of the hungry, exhausted men.

As the Marines were passing over a sharp ridge and descending to a piece of rare flat ground littered with many rock outcroppings, they immediately came under heavy fire from two Chinese platoons. Davis' men quickly attacked, eliminating the resistance. Several Marines were wounded and Davis called a halt so the casualties could be cared for. Many of the Marines sat down in the snow. It was easy for these exhausted men to fall asleep. Some had to be awakened and after a brief stop, the men pressed on.

Shortly before midnight, in total darkness, Davis realized the column was heading off course towards the road. He tried to pass the word to halt the march but his men had hoods over their heads and the wind was howling. Realizing that no one could hear, he started running up to the front of the line, through deep snow, with his radio operator and his runner. A lieutenant heading in the opposite direction, trying to locate his company commander, ran into Davis and knocked him over. He got up and continued to move ahead. Fortunately, the battalion had stopped because the point man had also figured out that he was leading them off course.

When Davis finally arrived at the front of the column, he was nearly spent and it took him awhile to get his breath. He knelt down in a depression in the ground and his runner placed a poncho over him so he could check his compass with a flashlight without being seen by the Chinese. Davis was exhausted and his mind was so numbed by the frigid air that when he stood up he forgot which direction he had decided on. He repeated the process but when he stood up again, his brain just wasn't working and he again forgot what he was doing. He tried a third time and he pointed in the direction he intended to travel and said, "This way."

Davis then turned to one of his company commanders and said, "Captain, am I making sense?"

The captain replied, "Yes, sir," but not without some doubt in his voice.

Davis looked at the point man and said, "All right, let's move out!"

Davis and his Marines continued to climb through the mountains. His battalion of 700 men stretched out over half a mile as it struggled through the frigid darkness. They had been climbing and moving for six straight hours. At 3:00 am, he realized that his troops had reached the limit of their endurance and he called a halt. He formed his men into a tight perimeter as quickly as he could. A fourth of the men stood watch; the others ate some dry, cold rations and slept on the snow covered ground. Captain Barber and F Company were less than a mile away.

HAGARU

Shortly before midnight, Smith and Bowser were reviewing all that went on that day. With operations running twenty-four hours a day at the command post, everyone lacked sleep. Personnel was so limited that there was little relief for staff officers. Smith and Bowser knew how hard it was on all the men. But Smith was optimistic.

Smith started the conversation, "Al, good things are starting to happen. With the airstrip open, we can fly out the wounded and can fly replacements in. Murray and Litzenberg are fighting their way south and Davis is on his way to rescue F Company. But I'm still worried about the soldiers on the east side of the reservoir. They have to fight their way over here!"

"Yes, General. We aren't hearing much from over there and that can't be a good sign," said Bowser.

"Al," Smith continued, "with all that's happened today, our first priority has to be to get this division from Hagaru to Koto-ri. We're going to have to fight our way down there. We're going to take all our trucks, tanks, heavy equipment, all our artillery and necessary supplies with us. Hopefully we can get all the wounded flown out in the next couple of days. If not, we'll take them with us. Anything we can't use, I want destroyed. I want nothing left for the Chinese. Al, give me a plan."

Smith's leadership style in commanding was to give clear, firm orders on "what" should be done but he never told his staff officers or unit commanders "how" it should be done. As the head of operations and planning for the division, it was Bowser's responsibility to craft the breakout plan. Smith and Bowser complimented each other and together they made a powerful team: one knew where he wanted to go; the other knew how to get there.

14

A DAY OF HEROICS

December 2

HAGARU

At 5:00 am, Smith was up and shaving when Bowser came charging through the door.

"Good morning, sir. Sorry to come barging in here like this."

"What's up?" replied the general.

"I've just received a report from the men out on the perimeter. The forward air controller with Task Force Faith, the Marine, Captain Stamford, came in early this morning. He said it was a massacre over there. Some of these soldiers from the east side of the reservoir have also been coming into our perimeter all night. Apparently, they came across the ice."

The general dried his face with a towel and said, "Get him up here right away and send someone to talk to those soldiers. I want to know exactly what happened over there."

Bowser left and fifteen minutes later returned to Smith's quarters with the captain in tow. When Stamford walked through the door he was dirty and his parka was stained with blood.

Smith immediately asked, "Captain, are you all right?"

Stamford noticed that the general was looking at his parka and replied, "Sir, I'm all right. It's not my blood."

Smith motioned for the two Marines to have a seat at his table and told his mess sergeant to bring some coffee. Bowser and Stamford took off their helmets and parkas and sat down. A minute later, the sergeant appeared with a pot of steaming coffee and three cups. Smith filled the cups and slid them over to Stamford and Bowser. Smith noticed that the Marine captain drank the coffee with his two shaking hands. His fingers were still purple from the cold.

Finally Smith looked at Stamford and said, "Captain what happened over there? Take your time."

Stamford spoke. "Sir, I'm not sure where to begin. I've been with Lieutenant Colonel Faith's battalion since the Inchon landing. My tactical air control party, including four enlisted Marines and I have been with him ever since."

Smith took the pipe out of his pocket and lit it and said, "Start with the night of the twenty-seventh."

"Sir, maybe I should back up a little bit. First of all, Task Force Faith had about 3,000 men but 700 of them were ROKs (Republic of Korea troops). They were untrained, undisciplined and most of them couldn't speak English. Whoever came up with the idea of mixing the ROKs with the American soldiers should be shot because they were worthless. The word is that most of the Korean troops were press-ganged off the streets, given little or no training and sent to war."

"Go on," said Smith.

"On the night of the twenty-seventh, we were north of the Pungnyuri Inlet and Lieutenant Colonel Faith put me with A Company because the next morning we were going to lead the attack north and he wanted me up front in case we needed air support. About 11:00 pm, I was bedded down in a bunker when I heard shooting. This didn't bother me at first because the ROKs were always firing at imaginary enemies. We had a poncho at the entrance to our bunker to keep out the snow. The poncho was pulled aside and here's the face of a Chinese soldier looking in. I blew his head off with my pistol but not before he threw a grenade in the bunker. It exploded on a sleeping bag and fragments wounded one of my men."

The Savior

Bowser asked, "How did the Chinese get inside the perimeter?"

Stamford replied with disgust in his voice, "I think the soldiers on watch were sleeping in their foxholes. It was utter chaos and several men were killed including the company commander and so I took over.

"When daylight came the Chinese backed off and four planes came in and they hit them with Napalm and strafing fire. This helped the situation on the ground."

"What about casualties?" asked Smith.

"We had eight killed and twenty wounded in A Company and there might have been another 100 wounded in the battalion.

"That night around midnight, the Chinese attacked again. By this time Colonel MacLean, who was now with Lieutenant Colonel Faith, decided to withdraw to south of the inlet to hook up with the rest of their regiment. We pulled out at 4:30 am on the twenty-ninth. All the trucks were unloaded of their supplies to make room for the wounded. The Chinese didn't follow us because they just went crazy plundering all that was left behind."

"Why weren't those supplies destroyed?" asked Smith.

"Sir, Colonel MacLean thought that if we burned the supplies it would alert the Chinese that we were breaking out. Anyway, this worked. When we were about to cross the inlet, Colonel MacLean thought he saw his troops coming from the south and he ran out on the ice. 'Those are my boys. I told you help was coming,' he kept shouting. But he was wrong. They were Chinese troops and they shot him several times and then they pulled him into the brush. We never saw him again. I assume he's dead. When we reached the rest of the regiment at the south side of the inlet, we found they were as shot up as we were. Many of the officers had been killed or wounded. Lieutenant Colonel Faith formed a new perimeter. In the afternoon, I called in airdrops of supplies. This helped a little. That night after midnight, the Chinese attacked again. There were more casualties but we held."

Stamford took a sip of coffee and continued. "Around noon, I called in more air strikes and more supplies were air dropped. That evening, Faith called a meeting of several of his key officers and told them, 'If we survive the night, we will break out for Hagaru in the morning. We can't wait for help that might not be coming.' We were hit hard that night but we were able to again fight them off.

"The next morning, there were more wounded. We were dangerously low on ammunition and even more discouraging, the weather was getting worse which meant the loss of air support and resupply by air. At 9:00 am, a Marine plane flew over the inlet and he radioed me that if the weather improved he would be back with a flight of Corsairs around noon. I gave this information to Lieutenant Colonel Faith."

Stamford stopped talking and was trying to find the right words.

Smith said, "Go on, Captain."

"Sir, this is when things started to go from bad to worse. Communications were awful. Some of the officers didn't even know we were breaking out and when the Chinese saw we were burning the supplies that we couldn't take with us, they came pouring out of the hills. The 600 wounded were put aboard the trucks as fast as possible. At 1:00 pm, the planes came overhead and the trucks took off down the road toward Hagaru."

Stamford took a sip of coffee and his hands were still trembling from cold and emotion.

"The lead pilot dropped a Napalm canister a split second too early. Most of the Napalm hit the Chinese but the tail end of it caught some of our people. Maybe a dozen soldiers were engulfed in flames. There was nothing the medics could do. This incident was a terrible blow to morale because the whole convoy had to drive by this disaster.

"A few minutes later, the Chinese attacked the soldiers at the front of the convoy and these men turned around and ran. Faith rushed forward and drew his .45 pistol. He pointed it at the retreating soldiers but he didn't fire. He yelled and screamed and told them to turn around and fight. The men finally did as ordered. Faith was sure a brave man, sir."

"I have no doubt of that," said General Smith. "Please go on."

"We had to deal with a blown out bridge, just north of Hill 1221, which slowed us down. As we detoured around it, we lost some trucks that broke through the frozen stream. Shortly before dark, we came to Hill 1221. The truck convoy could not move because the Chinese were pouring down heavy fire from the hills. They occupied the high ground on both sides of the road. Faith came across two ROKs who had tied themselves to the undercarriages of the trucks. I guess they thought if the trucks got through so would they. He ordered them to get out and join the fight. They refused and so he shot them both.

"As it got dark, the Chinese knew there would be no more air support and they charged down from the hills. Lieutenant Colonel Faith was badly wounded by a grenade and placed in a truck. His men tried to keep him propped up, in the front seat, even after he died, hoping the word wouldn't spread to the troops. I think it was at this time that the task force completely fell apart. I realized that things were hopeless and so I destroyed my radio and threw away my binoculars.

"When I came down from a small hill, I noticed the truck column had stopped. I walked down the road and found two broken trucks loaded with wounded, blocking the road. I found some walking wounded who helped me unload the injured soldiers and then reload them in other trucks. It took all the strength we had to push those trucks off the road. With the road cleared, I got the convoy moving again. Later in the evening, it was seven or eight o'clock, I moved about a half-mile down the road to see if it was clear of Chinese and roadblocks. I was with three or four soldiers and all of a sudden we were surrounded by Chinese. They captured us, took our weapons and told us to lie down along the side of the road. We did as we were told and then a few minutes later the guards got distracted and we took off running. I hid in some scrub pine and then slowly worked my way across the ice to Hagaru."

Stamford looked at General Smith and said, "Sir, with all the wounded and the dying it was like Custer's Last Stand."

Smith and Bowser were stunned by these comments. Finally the general asked, "Why did everything fall apart so quickly?"

Stamford was quiet for a moment and finally said, "Well, General, there was no communication between the units. Everyone was spread out all over the place. The radios didn't work and I think the soldiers were poorly trained, few had any combat experience and with many of the officers and senior enlisted men killed or wounded, no one came forward to take over.

"The Army high-command just threw these men to the wolves. They didn't have a chance!"

Smith said, "Thanks, Captain Stamford. You get some hot chow and sleep and report to Lieutenant Colonel Ridge. We're going to need you."

Stamford stood up and said, "Yes sir."

He put on his parka and helmet and walked out the door.

Bowser looked at Smith and said, "Jesus, General, what a disaster."

"I feel awful that we couldn't help these men," said Smith. "But we had no one to send them and we've been fighting for our own lives."

No sooner had Stamford left than Bowser got a message that General Almond was on his way to Hagaru.

When he told this news to General Smith, Smith said, "I wonder what he'll think about what we just heard from Stamford?"

"Who knows," replied Bowser. "We can only hope he finally realizes the mess that MacArthur's plans have gotten everyone into."

When Almond arrived, both Smith and Bowser fully recounted the fate of Task Force Faith. Almond's face blanched, he sat strangely silent on hearing the account of such a tragedy. He must have felt heartache, remorse and even guilt. After all, these were his men. But he said nothing.

They spent the rest of their time together discussing the plans to withdraw from Hagaru and the challenges involved.

After Almond left, Smith turned to Bowser and said, "At least he isn't telling me to burn my equipment anymore. Al, I think he finally is getting the picture of what needs to happen."

Later that morning, Lieutenant Colonel Olin L. Beall, the tough old salt commanding the 1st Motor Transport Battalion, was informed that Army survivors of Task Force Faith had been entering the perimeter at Hagaru all night. He was told that over 600 soldiers had passed through the lines, but that there were still a lot of men out there. Colonel Beall was fifty-two years old and had served in the Marine Corps for thirty-three years.

Beall had been a Marine in World War I but left the Corps after the war to pursue his real love, baseball. He had some success as a minor league pitcher before reenlisting in 1920. The second time around he was an officer. He saw combat in World War II as a major on Okinawa. He was a tough, no-nonsense officer of the "old breed" and his colorful personality endeared him to General Smith.

Beall, on his own initiative, with his driver, Private First Class Ralph Milton, drove the short distance to the south edge of Chosin Reservoir and chopped a hole in the ice to make certain that it was thick enough to support his jeep. He then roared north across the snow covered ice, driving up the middle of the reservoir. South of Hudong-ni he saw some men on the ice and drove over to investigate. He came under heavy sniper and automatic weapon fire when he started towards the shore. Leaving his jeep, he walked over to the men and found that many were wounded.

He said to one of the men, "Soldier, what outfit are you with?"

"We're part of the 31st Regiment, sir."

"What the hell happened to you men?" asked Beall.

"The Chinese attacked us five days ago and we tried to fight our way to Hagaru. Along the way, we had a lot of casualties and many of our officers were killed. The convoy then just fell apart. The Chinese had the

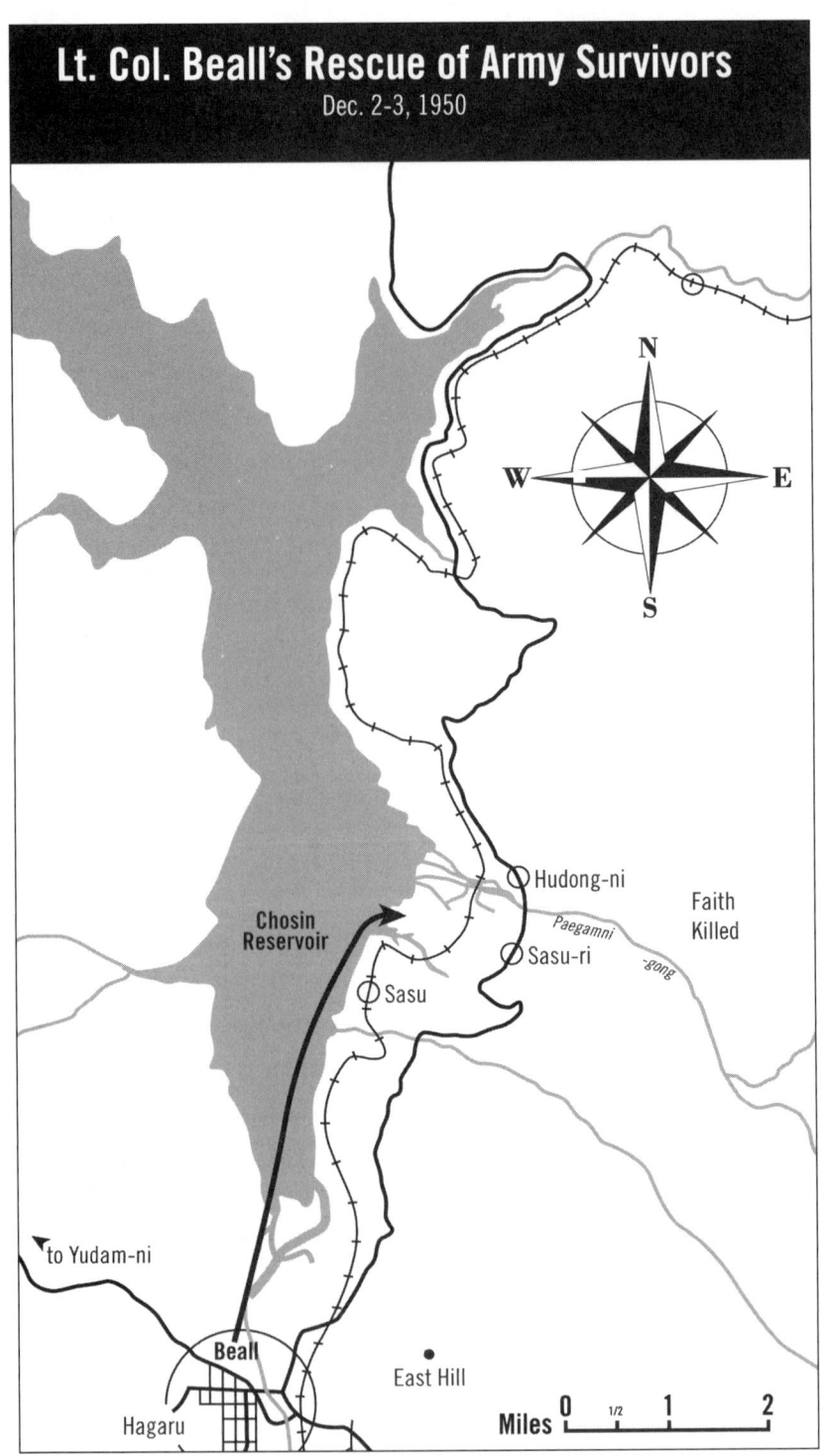

Lt. Col. Beall's Rescue of Army Survivors
Dec. 2-3, 1950

N

W E

S

Chosin
Reservoir

○ Hudong-ni

Paegamni

Faith
Killed

○ Sasu-ri

-gong

○ Sasu

to Yudam-ni

Beall

East Hill

Hagaru

Miles 0 1/2 1 2

The Savior

road blocked so our only option was to cross the ice on foot and hope we make it to Hagaru."

"Are there any more soldiers who survived?"

"Yes, sir. I think there's a lot of men hiding along the shore," replied the soldier.

Beall noticed that most of the soldiers were wounded and suffering terribly from the cold and hunger.

Beall barked out, "Private Milton, you load this jeep up with the people who can't walk and take them to the hospital. Then stop at the battalion and send an officer and trucks back here immediately."

Milton replied, "Yes sir."

The jeep was loaded and he took off, leaving Beall behind.

Beall then led the walking wounded down the ice towards Hagaru.

An hour later, Milton and Lieutenant Hunt of the motor battalion returned with several vehicles and drivers. Beall loaded the Army soldiers onto a truck, then he headed north across the ice with the remaining trucks in search of more survivors. He found plenty.

It became apparent to Beall that the Chinese were not firing at the wounded but only at the jeeps and trucks if they got too close to the shore. So he told his men to leave their weapons and vehicles out on the ice and walk in to help the soldiers. Parting a Marine from his weapon called for a real act of faith. But not one of Beall's men hesitated. Beall also noticed that if he and his men limped along like they were wounded, the Chinese would not shoot at them. The rescue operation went on all day. Beall did set up a machine gun on the ice to provide covering fire, if needed.

Some of the soldiers were wounded. Many had frozen hands and feet, and there were some with broken legs. There were those who had no boots because the Chinese had taken them and then set them free to die of the cold. All were disoriented from the sub-zero temperatures, lack of food and water. They wept openly on seeing their Marine rescuers.

Apparently the Chinese had not searched the marsh and brushland west of Hudong-ni so the soldiers who had escaped the convoy had been able to hide there undetected. When it started to get dark it was twenty-five below zero. Finding no more survivors, Beall returned to Hagaru. He and his small band of rescuers had saved the lives of 319 soldiers.

TOKTONG PASS

When daybreak came, Davis gave the order to move out. It took some time for the exhausted Marines to wake up and move. Snow covered their ponchos and sleeping bags, but the sergeants used a little "encouragement" and finally everyone was up and ready to go. Even these few hours of sleep had done much to revive their strength and spirits. The Marines were only a mile from Toktong Pass but they could not reach Captain Barber by radio. Was their relief too late, Davis wondered? He prayed it wasn't and pushed his men to press on, attacking to the south to rescue Barber.

Finally radio contact was made with Captain Barber. Davis let out a loud sigh of relief when he heard Barber's voice. He offered to send out a patrol to guide in the battalion but Davis declined. "Save your men and your ammo. We are on our way," Davis told him. Just before noon the forward elements of his battalion reached F Company. The Marines stopped because they couldn't believe what they were seeing: F Company was surrounded by a sea of frozen Chinese bodies. In places the bodies were stacked several feet high to form barricades. Blood stains covered the white Chinese uniforms.

After five days and nights of intense fighting, Captain Barber's company had suffered 118 casualties: twenty-six killed in action, three missing and eighty-nine wounded; a casualty rate of fifty percent. Some men were wounded two or three different times. Six of the seven officers including Barber were wounded. Most of the unwounded men were suffering from frostbite and digestive ills. The beleaguered company had held this strategic position against unbelievable odds. Their survival defied rational explanation.

As he entered the perimeter, Davis saw Captain Barber being carried out of his command tent on a stretcher. Using a tree branch as a crutch, the captain insisted on standing up to shake hands with Lieutenant Colonel Davis. It was such an emotional moment for both men that tears filled their eyes and neither could speak.

Finally, Davis managed to say, "Captain, I don't know how you held on for so long. You and your Marines have done one hell of a job."

Barber just stared at Davis and said in a choked-up voice, "Thank you, sir. I'm very proud of my men. They gave it their all! We're damn glad to see you."

With his five day stand, Bill Barber had held the pass against amazing odds and protected the only route that Litzenberg and Murray would have to take to bring their men home.

Davis had twenty-two of his men wounded during the daring rescue and they were carried into the perimeter on stretchers. While looking after these men, the regimental surgeon, who had volunteered to go with Davis, was killed instantly by a sniper's bullet. There were no other Marines killed, although two men went berserk and had to be placed in improvised strait jackets. They died before they could be evacuated.

Davis's battalion spent the rest of the day enlarging and reinforcing the perimeter and pushing the remaining Chinese off Toktong Pass.

The 5th and 7th Marines were still fighting their way south and would not reach Toktong Pass until the following morning.

HAGARU

Other than attacks by the Chinese, the biggest problem facing the garrison at Hagaru was the evacuation of the wounded. All day planes landed and took off with the casualties. After the last plane of the day had taken off, Dr. Hering reported into General Smith in his command tent.

"Sir, we have a problem."

"What's that?" asked Smith.

"A bunch of the soldiers who came in here last night from Task Force Faith were not really wounded but they got aboard the planes and were flown out of here."

"How can that be?" replied Smith.

"We flew out 919 men today. This morning, we had 450 men in the medical facilities but at the end of the day 260 remained. It appears that a significant number of men who were not casualties flew out of here; 729 men to be exact."

"I don't understand," reacted Smith.

Smith was getting mad and demanded, "How could something like this happen? Those planes are here to take out our wounded and you're telling me the wounded are still here? These soldiers have pushed their way on planes with no regard for the welfare of the wounded?"

"Sir, the Air Force had sent up an 'Evacuation Officer' to supervise the loading of planes. I just assumed that he was responsible for screening casualties that got on the planes. But he told me that was not his responsibility. He did however become suspicious that many of these men were not casualties since they were walking and talking as they boarded the planes and he finally reported it to me."

Smith exploded, "I will not tolerate such cowardly behavior! This is a disgrace! Doctor, you put one of your men in charge of this and I want these casualties tightly screened. These pilots are risking their lives to bring out the wounded. And there are lots of real wounded waiting to be evacuated."

"Yes, sir," responded Hering. "I couldn't agree more."

Hering selected one of his own surgeons, Lieutenant Commander Lessender, for the task. Lessender's feet were painfully frozen but he had refused to be flown out. He was just the man to put in charge of who would be evacuated and who wouldn't. There would be no easy exit from

Hagaru. Lessender immediately devised a protocol for determining the severity of frostbite. He would decide who would walk, who would ride and who would fly. Others were treated and sent back to duty.

A half hour after Dr. Hering left, Colonel Bowser came into the command tent.

"Sir, the men have been putting up the warming tents and stoves for the arrival of the 5th and 7th Marines. We're also setting up galleys so we can feed all these men. We hope to have all this completed by midday tomorrow."

"Al," Smith said, "when do you think the lead elements will get here?"

"Sometime tomorrow afternoon," replied Bowser.

"I just want to make sure we can handle all these men."

"Sir, we'll be ready."

Shortly after dark, Lieutenant Colonel Beall reported into General Smith.

"Good evening, sir," greeted Beall. "You wanted to see me?"

"Colonel, I heard you've had quite a day. Have a seat. Tell me what happened."

Smith was surprised how beaten down Beall looked but then he thought everyone is starting to look like this. Like Smith, Beall was one of the older Marines at Hagaru. Beall didn't drink or smoke and Smith knew he kept physically fit, but tonight it was clear that Beall was feeling his age.

"How many soldiers did you bring in?"

"Over 300, sir."

"What condition were they in?" asked Smith.

"It was a pitiful looking bunch, sir. Some of the men were limping along the ice. Some were crawling because they had no boots. Some were wounded and some had broken bones. Others were totally disoriented and just walking around in circles. They had no idea where they were going. Most of these soldiers had lost their weapons or abandoned them and the sub-zero weather has damn near killed these men. They kept coming out of their frozen hiding places along the shoreline. When my men got close, they looked like ghosts. They were snow covered and damn near frozen."

"Do you think there are any more soldiers out there?"

"I do and with your permission I'd like to go back tomorrow. And if I can get up on the shore, I'd like to check out those trucks. There might be some of the wounded men still alive in there."

"Colonel, that was quite a rescue operation. Well done. Let your men know they are a credit to the 1st Marine Division. You be careful out there tomorrow."

"Yes, sir." Beall stood up and left.

That night remained quiet at Hagaru and Koto-ri but not so for the 5th and 7th Marines as they continued to slowly fight their way down the road toward Hagaru.

15

FROSTBITE

December 3

TOKTONG PASS

At 6:30 am, Lieutenant Colonel Davis's rifle companies swept the Chinese off the surrounding hills that led south. When this happened, the enemy retreated north up the road right into Litzenberg's Marines coming south. These Chinese were now trapped. Artillery was called in from Hagaru and the enemy was mowed down by the howitzers. Then the Marine Corsairs arrived on the scene and they strafed and killed anyone who was left standing.

Later in the morning, Barber's men saw the Marines from Yudam-ni marching toward them. They were overwhelmed. Some of the men had tears streaming down their cheeks. This is what they had fought and died for, to keep the pass open for these Marines.

When Litzenberg arrived at Toktong Pass, he assigned an officer to stop each truck at the bottom of the hill and try to cram one more wounded Marine from Barber's Company on board. The problem was, Litzenberg's trucks were full of 1,000 wounded and frostbite cases. All of the wounded men of F Company were eventually loaded onto the trucks and some were tied to hoods and bumpers because of lack of space in the vehicles. The Marines who were killed at Toktong Pass had to be buried there. Regrettably with so many wounded, there was just no room for the dead. The priority had to be saving the living.

The road was now open to the south and Ray Davis's battalion led the way. They were followed by the long motor column as it slowly wound its way to Hagaru.

HAGARU

In the morning after the staff briefing, Smith was jeeped over to E Medical Battalion to confer with Dr. Hering. The welfare and condition of his men was constantly on his mind. His staff knew that casualties were always first on his daily agenda. Nearly every night, his notebook was filled with numbers of killed in action, wounded in action and missing in action. Each number represented a Marine under his command. He personally felt responsible for each entry in his journal.

Smith entered the one-story building which housed most of Hagaru's casualties. The scene of human suffering was almost unbearable. This was the ultimate price of war. The stringent smell of antiseptic mixed with the foul odor of unwashed Marines was overwhelming. Sounds of the wounded and dying men, heavily sedated with morphine, provided a haunting noise.

Every inch of the building was occupied by stretchers; stretchers on floors, stretchers stacked three high on racks. Down the center of the medical building was a row of sawhorses supporting wooden platforms on which the most seriously wounded were being operated on in the most primitive of conditions. Looking around, Smith thought to himself that except for the bitter cold, this could be a scene from the Civil War.

General Smith watched intently as a Navy corpsman was examining a young Marine's feet. Smith walked closer to the Marine and as he did, he noticed the discoloration of his toes. Frostbite, he thought, was hard for even the non-medical to miss.

The general found Dr. Hering conferring with another doctor. Smith waited until they had finished. Dr. Hering walked over to General Smith.

"Good morning, General. How may I help you?"

"Doctor," started Smith, "I've been watching some of your corpsmen tend to our Marines just now and I've been reading the daily reports: more and more cases of frostbite. I grew up in California and I've spent most of my life and career in warm, even tropical climates. I once served at a duty station in Iceland but the temperatures there were mild compared to what we're experiencing here. These constant sub-zero temperatures and what it's doing to our troops is not something I'm familiar with. Can you give me a quick lesson on frostbite?"

Dr. Hering leaned back against a table. He looked exhausted. Smith knew that the stream of dead and wounded must be taking a terrible toll mentally as well as physically on these doctors.

"General Smith," the surgeon started, "we have lost count of how many frostbite cases we've had in the last several days and the numbers keep getting higher. This unrelenting cold is taking its toll on the men's bodies. I'll try to give you a simple explanation.

"When skin is subjected to prolonged exposure to very cold temperatures, frostbite will occur. Many times, the men don't even know that anything is wrong because the affected areas such as the face, fingers and toes are numb and lose all feeling.

"When the frozen skin becomes numb and the area gets hard and waxy looking, skin discoloration starts and eventually the skin turns black. Depending on how long and how deeply the tissue was frozen will determine the permanence of the injury. In many cases we've seen recently, the blood flow to the frozen areas has stopped. Skin, tissue, blood vessels, muscles and nerves become permanently damaged. The frozen tissue dies and amputation is necessary.

"Obviously, frostbite is a danger whenever the temperature stays below freezing but the big difference here and the reason we're experiencing so many cases is the temperature extremes and the prolonged exposure to it. With temperatures plunging to twenty-five or thirty degrees below zero, it accelerates the freezing process. And that General, in a nutshell, is frostbite. It affects everyone, all the time. Man was just not made to live and fight in this extreme cold."

General Smith had remained quiet and attentive during the doctor's explanation. When Dr. Hering finished, Smith said, "Thank you, Doctor. I have a better understanding of what we're dealing with now. It's important to understand the problem so that we can try and fix it."

Hering added, "General, most of the frostbite we're seeing is the result of those damn shoe-pacs. The rubber bottoms on these boots don't breathe and so when these men walk or climb a hill, their feet start to sweat. The sweat goes through their woolen boot socks and then the felt inner-soles. When they stop, everything freezes including their wet feet.

"As you know, every man has been issued an extra pair of inner-soles and socks and they've been instructed that when they become wet, they should replace them with the dry set. The wet set should be placed under their parkas, next to their skin, so they dry out. But with the constant fighting and the bitter cold, this just isn't happening. They just don't have the luxury of frequent sock changes."

"Doctor," said Smith, "I assume that many of the men in here are the Army soldiers that Colonel Beall rescued yesterday."

"That's correct sir," responded Hering, "but we also have a good number of Marines in here. We are already starting to fly them out and we hope to have them all evacuated before the Yudam-ni Marines arrive. The planes are flying despite the snow."

"Doctor," replied Smith, "I know that you are doing the best job you can with what you've got. You have my sincere respect and gratitude."

As Smith left the field hospital, he wondered how the human body could tolerate such extremes. He remembered that in World War II on Peleliu that the temperature would sometimes climb to 110 degrees. Men dropped like flies from heat exhaustion. It was like fighting on the surface of a wood stove. Now here, we're dealing with temperatures that have dropped to thirty degrees below zero. That's a 140 degree swing!

The Marines at Hagaru were busy all day setting up warming tents and galleys for the arrival of the 5th and 7th Marines. The number of Marines at Hagaru would triple in size and the medical facilities would be taxed to their limit. Feeding and housing the new arrivals would be a monumental task. Airdrops of supplies continued through the day.

At 4:30 pm, General Smith sent out a relief force consisting of tanks and the Royal Marine Commandos to drive the Chinese off the hill nearest the entrance to the perimeter and to clear the road. He didn't want the enemy to interfere with the anticipated arrival of the 5th and 7th Marines. These men had come too far to have to fight all the way to the perimeter.

By 7:15 pm, General Smith was in the command post when he was told that the 7th Marines were coming in. He got in his jeep and was driven to the north end of the perimeter where a large crowd of Marines had gathered to greet the arriving Yudam-ni Marines. Although it was dark, the starlight reflected off the snow and you could see and hear the convoy coming down the road. They halted 600 yards outside the perimeter. Ray Davis ordered his exhausted Marines to form up and he said, "Look sharp. We're going in like United States Marines."

A sergeant stepped out of the line and started calling cadence in that halting, melodic sound. The Marines stood tall and marched in with shoe-pacs pounding the frozen ground in a slow, steady rhythm. These Marines were filthy. Washing and shaving hadn't been an option in Yudam-ni. Their hair was matted and their parkas were covered with an assortment of filth: mucous, blood, urine, signs of dysentery, and stains from C-rations. They were cold, hungry and utterly exhausted. They hadn't felt their feet for hours. Seldom has the human body been so savagely punished.

Next came the walking wounded and then the jeeps and trucks loaded with those Marines who could not walk. Most of the vehicles were riddled with bullet holes and many had their windshields shot out. The wounded men were packed in the back of the trucks and there were some who were roped across the hoods of the jeeps or tied to the fenders.

The Marines at Hagaru were there to greet their comrades from Yudam-ni. As they heard the cadence they began to clap and cheer. Men who had been facing death for days began to cry at the sight of Davis' incoming troops. Their tears froze in the Arctic night air and they surged forward to help their brothers.

It was an emotional reunion for everyone. For Smith it was because he finally had most of his division together. With the exception of two of Chesty Puller's battalions, which were holding Koto-ri and Chinhung-ni, his family of Marines was reunited. Now there was no doubt in Smith's mind that Hagaru would hold.

For the incoming Marines, Hagaru appeared to be an oasis of safety after all they had gone through at Yudam-ni. They soon found out that Hagaru had been cut off and isolated just like Yudam-ni. And for the Marines at Hagaru, their weakly defended perimeter would now be heavily reinforced and that gave them a great sense of relief.

They all knew they still had to fight their way south but for now there was safety in numbers. Their brothers had made it back and so would they. Morale soared!

All casualties were given as much care as the overburdened medical companies could provide. Guides were on hand to lead the incoming Marines to preassigned tents. Vast quantities of coffee, hot stew, pancakes and syrup were provided to the cold and ravenous troops. The hot food brought the men back to life. Men who had barely survived actually found themselves able to laugh again.

When Smith returned to his command post, Lieutenant Colonel Beall was waiting for him.

Smith started, "Colonel, I'm sure you've heard that the Marines from Yudam-ni are starting to come through the perimeter. I am starting to feel a lot better about our situation. How about your rescue operations? Did you find any more soldiers out on the ice?"

"Yes I did, sir."

"How many men did you bring in?"

Beall said, "Sir, I left at about 8:00 this morning and there were reports that there were more men out on the reservoir. My driver, PFC Milton and Corporal Howard and I drove out on the ice and we found four seriously wounded soldiers who were almost frozen. As we approached them, we came under machine gun and rifle fire. But out of the blue the planes came in and saved our lives by knocking out the machine guns. When the Chinese started to run, the planes just tore them apart."

"What condition were these soldiers in?"

"Sir, these men had been taken prisoner and they were put in a house. But then the Chinese brought them down to the reservoir and then shot at them. They all had leg wounds and were unable to walk. Two of the soldiers had both legs broken, one had one leg broken, and the other had flesh wounds. These men were taken back to Hagaru. Then PFC Milton drove me over to where the convoy ended up. You could see the trucks from the shore. I told Milton to stay in the jeep and I walked up a slight hill to the road. I looked in every truck, jeep and trailer. But, I struck out, sir. They were all dead."

"Where were the Chinese?"

"Sir, I didn't see any and if there were any of them around, they left me alone."

"Colonel," said Smith, "how many bodies were there?"

"Sir, this is just an estimate, but 300 would be a pretty good guess. There were frozen bodies in the trucks and all over the area where they had been abandoned by their own men and the trucks had been destroyed."

"General, as far as I can tell, the task force lost everything: trucks, jeeps, artillery pieces, mortars and machine guns. Sir, they didn't bring anything out. It was all lost. And that's just the equipment. These soldiers tell me they started with about 2,500 men. They estimate half of these are dead or missing and less than half of the living are in any condition to fight."

"That was a real tragedy," said Smith. "If it wasn't for you, those soldiers would have died out on the ice."

Beall stood up and looked at Smith and said, "In all my years in the Marine Corps, I've never seen anything like this. Ever. One other thing sir, Colonel Litzenberg will be arriving some time tonight and I've got room in my tent and if it's all right with you, he can stay with me. Sir, the colonel is an old friend."

"Certainly," replied Smith. "I will make sure to tell him when he gets here."

Beall said, "Good night, sir," and walked out of the command tent.

Half an hour later, Litzenberg reported in to General Smith.

"Litz, it's great to see you. Welcome to Hagaru," greeted Smith. He had never seen Litzenberg in such a state. He was dirty, unshaven but still full of fight.

"General, I can't tell you how good it is to be here," answered Colonel Litzenberg. "We had a rough time of it up at Yudam-ni and getting down here was no picnic."

The colonel reached inside his parka and pulled out a pack of cigarettes. He lit one and took a long drag. Fatigue showed on every line of his face.

"Litz, how are your men doing?" asked Smith.

"They've been through hell, sir. They're just plain worn out and a lot of them need medical care, but making it to Hagaru is a big boost to their morale. General, I've never commanded a tougher bunch of men."

"Colonel Beall was just in here and he requested that you bunk in his quarters tonight," said General Smith.

"Thank you, sir. That's very nice of him."

"Corporal," motioned Smith, "drive Colonel Litzenberg to Colonel Beall's quarters."

The Savior

As Litzenberg was leaving the command post, he looked at General Smith and said, "I sure can use a good night's sleep."

"You've earned it," said Smith.

Later that night, Lieutenant Colonel Murray reported to General Smith. When the colonel walked into the command tent, Smith was stunned by how Murray looked. The general had not seen him for two weeks. He had aged years in that short time. His face was gaunt, his eyes sunken and he had lost weight. The stubble of beard made him look older than his years. His hair was matted and his combat clothes were filthy.

"Ray," said Smith, "why don't you stay with me in my quarters. I've got an extra cot in there for you. We'll get some hot water for you and you can clean up a little."

"Thank you, sir. I'd appreciate that. I'd make a pretty ripe roommate if I couldn't clean up."

They walked the short distance to Smith's house and Smith said, "Ray, how about something to eat?"

"Sir, that would be wonderful. I haven't had a hot meal since Thanksgiving Day."

Murray took his pack off his shoulder, removed his parka and helmet and put them on the empty cot.

Smith told his mess sergeant to prepare something for Colonel Murray to eat and to get some hot water for washing.

Smith and Murray sat down at a table and Smith said, "Ray, tell me what happened up there."

Murray placed his hands on the table and said, "Sir, I've got to tell you it was like being trapped in a nightmare when you can't wake up. On the night of the twenty-seventh, the Chinese hit us hard and the 5th and 7th Marines suffered many casualties. The next morning, without any orders to the contrary, I decided to postpone the attack west. It would have been just plain idiotic to continue that attack. Colonel Litzenberg then approached me and he told me he wanted to establish a joint command. For me this was a great compliment and by the end of the day, I moved into his command post which was a small stone house. I was very grateful that he treated me as an equal. After all, Colonel Litzenberg is forty-seven years old and I'm a thirty-seven-year-old lieutenant colonel. The first thing we did was combine our staffs and reorganize."

Smith lit his pipe, smiled his approval, and said, "Go on."

"Both our regiments had taken a terrible beating on the evening of the twenty-seventh. I strongly believe that the Chinese suffered many more casualties than we did but the problem was they had a hell of a lot more men. There was just no end to the men they threw at us. It looked like all of China was out there. Anyway, in the morning we consolidated our position and tightened up the perimeter. For the next few days the Chinese would take the high ground at night and we'd take it back during the day."

"What about your casualties?" asked Smith.

"That was a huge problem. We had a very difficult time handling all the wounded. The docs and the corpsmen did an unbelievable job but we had so many wounded that we couldn't put them all in the aid stations. So the medical people laid them on the ground outside on straw. Then the corpsmen covered them with tarps and they would frequently walk by and brush the fresh snow off their faces of those men who were so seriously wounded that they couldn't do it themselves. It was easy to tell who was alive and who was dead. The faces of the dead were covered with snow."

It was obvious to Smith that Murray was having a difficult time talking about this. The mess sergeant walked in and put a huge plate of beef stew in front of Murray. Then he brought him a large cup of black coffee.

Smith said, "Eat up, Ray!"

"Just need one thing, General." Murray stood and walked over to his pack where he removed a bottle of Tabasco sauce. When he returned to the table, he doused the stew with the fiery red liquid.

Smith didn't say anything. He just watched him eat and thought, those Texans sure like their hot sauce. When Murray was finished he said, "That was the finest stew I have ever eaten."

"Glad it hit the spot. Continue with your report, Ray."

"Luckily General, the Chinese made two big mistakes. The first was when my 5th Marines attacked west from Yudam-ni on the twenty-seventh. They should have left us alone for a day or two. We would have been strung out along that road and separated from the 7th Marines. They would have cut us to pieces. But they jumped the gun and attacked us, forcing us to pull back to Yudam-ni.

"Then I believe the Chinese made their second big mistake. They had us surrounded and they were attacking us everywhere. If they had concentrated their soldiers on the road just south of Yudam-ni, we would never have broken out of there. But by surrounding us they weakened their strength. It's fourteen miles from Yudam-ni to Hagaru. It took us fifty-nine hours to fight our way down here. When everyone gets here we'll have about 1,500 casualties: 1,000 from wounds and 500 from frostbite. When this war started, I never heard of Yudam-ni. Now I'll never forget it."

Murray took a sip of coffee and then said, "Sir, I can tell you this with absolute certainty: whatever lies ahead of us can't be worse than what we've left behind. I've never told anyone this but I didn't think we'd get out of there. My first combat was on Guadalcanal in January of 1943. I commanded the 2nd Battalion, 6th Marines. Fighting the Japanese in that stinking jungle was absolutely brutal. The heat and then the malaria just drained you. We got very little sleep and you lost weight at an alarming rate. Your belly just ached with hunger. You were always wet from rain or sweat. At the time, I thought that nothing could be worse. Well sir, Guadalcanal was a picnic compared to this. Sir, I have never been so proud to be a Marine. Every day, every hour you would see or hear about some amazing feat of bravery. Men who put themselves in the line of fire to provide cover for their troops. Men who bought time so others could withdraw.

"General Smith, I heard what happened to Faith and the soldiers in his command. I warned them when they relieved me on the east side of the reservoir not to get too strung out. But I'm sure they didn't have any choice when General Almond ordered them to get to the Yalu River in a hurry."

Smith stood up and said, "Ray, you and your men should be proud of yourselves." Then he said, "It's getting late. Your hot water is ready. You clean up and get some sleep. I'll be back."

Smith left and walked to his command tent. He felt a great sense of relief that Murray and Litzenberg had made it to Hagaru. The pieces were starting to fall into place.

Later that night, Dr. Hering walked into the command tent and spoke with Smith.

"Good evening, General."

"Good evening, Doctor. How did you make out today?"

"Sir, we flew out 700 casualties today."

"Doctor," asked Smith, "are you sure that all these men were injured? I don't want any of these jokers getting on these planes."

"Yes, sir. I can assure you that no man is being evacuated unless he is a casualty; either from wounds or frostbite."

"Doctor, are you going to be able to handle the Marine casualties from Yudam-ni?"

"General," replied Hering, "my men are ready and we are looking at their people as they come in. We are screening them, treating them and tomorrow as soon as the strip opens, we'll fly out the most serious cases first. I just hope the weather continues to cooperate."

"Thank you, Doctor," said Smith. "You and your men are doing a remarkable job."

"I'll be honest with you, sir. We are fighting two battles here; one against the Chinese and one against the cold and I sometimes wonder which one is worse. We are doing our best. Well sir, I better get back; it's going to be a long night."

When Smith returned to his house Murray was fast asleep.

The Chinese did not attack Hagaru that night.

GENERAL SUNG

Chinese casualties were horrific and they were rapidly mounting. The bitter cold would fell almost as many men as would enemy fire. Soldiers would die from exposure, frozen in their fox holes. Because there was little or no medical help available to the Chinese, a wounded soldier was left alone to drag himself north toward home while constantly hoping not to freeze to death along the way. In most cases, a wound to a Chinese soldier was a death sentence. Many Chinese soldiers carried Benzedrine pills and opium to self-medicate.

The Chinese lacked tents, trucks and resupply lines. Their soldiers were generally illiterate peasants conscripted into service. And yet by sheer numbers they were pushing the Americans out of North Korea. But they also were not defeating the enemy they sought to destroy. They repeatedly failed to press their advantage. Whether due to poor leadership or supply issues, when the Chinese succeeded in breaking through American ranks, they would often go off in a frenzy of looting for food and warm clothing rather than pressing the battle. They also would repeatedly attack in the exact same location at the same time, allowing the Marines to be prepared for the attack. They seemed incapable of sustaining an operation. Vicious attacks were sometimes followed by a quiet day.

With General Sung's optimism sinking, he continued to ruthlessly throw wave after wave of his starving, freezing men against the deadly fire of the Marines. Much to his dismay, the men of the 1st Marine Division could not be defeated. Chairman Mao's assessment that the 1st Marine Division had the highest combat effectiveness had proven true. General Smith had proved a most formidable opponent.

16

THE FORTUNES OF WAR

December 4

AMERICA

Unbeknownst to General Smith and his Marines at Hagaru, the newspapers in America were predicting that because the 1st Marine Division was surrounded by tens of thousands of Chinese soldiers, they were in danger of being annihilated. The famous newspaper and radio commentator, Walter Winchell, in a weekly broadcast said, "If you have a father, brother or son in the 1st Marine Division, pray for him tonight." Even though the mood in America was very pessimistic about the Marines' chances for survival that was never the feeling of the men in Hagaru.

HAGARU

At first light, Smith again sent out the British commandos and tanks to help protect the remaining members of the 5th and 7th Marines as they entered the perimeter of Hagaru. He had come to like and value these Royal Marines. The last U.S. Marines would enter the base at 2:00 pm and were welcomed by all of Hagaru.

Low clouds hung over Hagaru in the early morning hours. However, at 9:00 am the weather started clearing and air evacuations of the wounded and airdrops to resupply started and would continue throughout the day.

The Savior

Thanks to General Smith's keen foresight in establishing a landing strip at Hagaru and the magic worked by Partridge and his engineers, desperately needed items for their breakout were flown in. As the 5th and 7th Marine Regiments joined the Marines defending Hagaru, the need for supplies became critical. The men at Hagaru would need the following items: 1.5 million rounds of small arms ammunition, 37,000 mortar and artillery shells, 10,000 hand grenades and 46,000 C rations. Thousands of gallons of fuel was needed to keep all the vehicles running twenty-four hours a day in the extreme cold.

Colonel Bowser stood outside watching the magnificent sight of hundreds of parachutes with their air drop containers as they drifted into the Hagaru perimeter. It was a vision that warmed the cockles of an operations officer's heart.

Just before noon, Bowser received a radio dispatch from air command at Yonpo that they were running dangerously low on parachutes and air drop containers. Future airdrops would be affected. Colonel Bowser immediately issued a directive that all parachutes be rounded up from Hagaru and returned to the supply officer at the airfield for return to Yonpo air command.

Back in the command tent while Bowser was explaining all this to Smith, a sergeant interrupted and said, "Excuse me, General. There's a call for you from X Corps."

Smith got on the radio and said, "Yes?"

"Sir," answered the unidentified voice, "it has been reported that you are flying out the dead along with the wounded."

"That's correct," replied Smith.

The voice on the other end continued, "General, we feel this is an unnecessary use of space on the planes. It could be preventing evacuation of some of the wounded."

Smith angrily replied, "The wounded will always have first priority, but if we have the room we will continue to fly out the dead. My men have risked their lives to recover the bodies of their fellow Marines. And by God, if there is available space, we're flying them out." Then Smith just hung up.

X Corps did not call back.

Later that morning, Lieutenant Colonel Partridge reported in to General Smith. "Sir," started Partridge, "we have a very serious problem."

"What now?"

"Sir, intelligence reports that the Chinese blew the bridge at Funchilin Pass where the power generating plant is." Pulling out a map, Partridge pointed, "As you can see, the bridge is three miles south of Koto-ri and neither trucks nor tanks can cross the bridge now. The road here runs along the edge of a 1,000 foot cliff. If we don't get this fixed, the troops could still walk out but we couldn't take any of our equipment. I have an idea on how it can be repaired with some treadway bridge sections, but I want to check this out myself. I just wanted you to be aware of the problem. I'll have my recommendations to you by the morning."

"John, we have to get our vehicles across that bridge," insisted General Smith. "Not only will they be carrying wounded but they will be essential to protect the troops and we sure don't want them to end up in enemy hands to be used against us."

"Understood, sir," said the engineer.

Partridge left. As usual, he had to find a solution.

As the last Marines were coming into Hagaru, Smith held a meeting with his senior staff, his two regimental commanders and Dr. Hering who was also present. Smith had a difficult decision to make. Should the Marines break out for Koto-ri the following morning or should they delay it another day until the sixth of December?

Smith stood up and spoke.

"Gentlemen, X Corps is pushing us to break out of Hagaru tomorrow. The latest intelligence reports indicate that enemy reinforcements are arriving in the Chosin Reservoir area. If we delay an extra day it would give the Chinese time to add more troops to block all movement south. The problem is that our troops are worn out. Should we move out tomorrow or should we wait a day?"

Dr. Hering spoke first. "General, from a medical perspective, I think it would be foolish to leave tomorrow. The Marines coming in from Yudam-ni have approximately 1,500 casualties and a third of those are frostbite. We looked at a lot of these men throughout the night and into the morning. They are really in bad shape. These men are exhausted. They need food, sleep and medical care. It is my estimation that if you leave tomorrow, half of these men would never make it. In their condition, it's a miracle that they got here at all. They are asleep on their feet as it is now. And one other thing, we have so many casualties to evacuate that it would be impossible to get them all out today or even tomorrow."

Smith looked at Colonel Litzenberg and said, "What do you think?"

He replied, "Sir, I realize if we delay a day we give the Chinese a chance to strengthen their forces but I think we must take that chance. I strongly agree with Dr. Hering. My men will do a lot better with some rest and food. An extra day will make a big difference. It sounds like we are going to do a lot more fighting on the way south and a little more time to recover will help."

Smith said, "Ray, what is your opinion?"

Murray replied, "I agree with Colonel Litzenberg and Dr. Hering; let's wait another day."

The general could see Colonel Bowser nod in the affirmative.

Smith said, "Alright, we will delay moving to Koto-ri until the sixth. Gentlemen, take care of your men and see that they get proper medical care, plenty of food and rest. I'll let X Corps know our plans."

Just then, they heard the whirl of an incoming helicopter. The phone rang, Colonel Bowser answered. He listened and after hanging up he said to Smith, "Sir, it was General Almond. He just landed and he wants to talk to you, Colonel Litzenberg, Colonel Murray and Colonel Beall. I'll get Colonel Beall up here right away."

A few minutes later, Major General Almond walked into the command tent. Everyone stood and greeted Almond.

"Good afternoon, General," they said in unison.

"Good afternoon, gentlemen," Almond continued. "General Smith, I've made arrangements to have B-17s and B-29s escort you all the way back to Hamhung. Their bombs will clear a path for you all the way to the sea."

Almond apparently still doesn't realize how tough this is going to be, Smith thought to himself. More than a little bit of sarcasm could be heard in Smith's reply, "General, I think there will be sufficient fighting for all of us."

Lieutenant Colonel Beall entered the command tent. After greetings were exchanged between Beall and General Almond, the X Corps commander said, "General Smith, Colonels Litzenberg, Murray and Beall, please line up. It is my honor to award you the Distinguished Service Cross for your actions at the Chosin Reservoir. The citation reads as follows: 'For extraordinary heroism in connection with military operations against an armed enemy during the period twenty-nine November to four December, 1950. Your actions contributed materially to the successful breakthrough of United Nations Forces in the Chosin area.' Congratulations."

Smith thought that this was neither the time nor the place for an awards ceremony. The Distinguished Service Cross was the Army's second highest medal and it was a great honor to receive it. But Smith knew that there was still much more fighting ahead. In a way, Smith believed, Almond is more like MacArthur than MacArthur in the way he likes to pass out medals.

Almond reached in his pocket and said, "I only have one medal. Who should get it?"

They all agreed that it should go to Lieutenant Colonel Beall. As Almond was pinning the medal on Beall, Smith wondered why Almond would come all the way up here to decorate and only bring one medal. Smith just shook his head.

Almond continued, "Gentlemen let's sit down and talk."

He looked at General Smith and asked, "What are the plans for the breakout?"

"We decided not to leave until early morning on the sixth and the breakout plan is not yet completed," answered General Smith.

"Why not leave tomorrow?" asked Almond.

"The Marines from Yudam-ni are exhausted; they have been in constant combat. They've had little or no sleep for eight straight days. They've had no hot chow and many of the men have dysentery from eating frozen rations. A lot of these Marines have frostbitten feet and hands from this sub-zero weather. General Almond, these men are in bad shape! Besides, it's going to take a couple of days to fly out all these casualties."

Almond reluctantly agreed and wished them well. He seemed less inclined to question Smith's battle plans. He was then whisked away by air to Koto-ri to present a Distinguished Service Cross to Colonel Chesty Puller.

Shortly before dark, Dr. Hering returned to the command tent where he found General Smith conferring with Colonel Bowser.

"What's the status of the wounded evacuations?" asked Smith.

"General," started the doctor, "we've airlifted about 1,000 casualties out of here today."

"Dr. Hering, that's amazing," said Smith. "How many more do you think we have left?"

"A good guess would be about 1,400 to go," replied Hering.

"Doctor, do you think we can fly all these remaining casualties out in two days?" asked General Smith.

"There is a chance, but only if the weather holds," replied Dr. Hering.

"Well, we'll have to pray for good weather. Good work, Doctor. Please make sure your men know how much their services are appreciated. You do unbelievable work, often in the line of fire. I remember reading the casualty figures on Iwo Jima just before we attacked Okinawa. I was sickened when I learned that twenty-three surgeons and 827 corpsmen were killed or wounded in the battle. Later four corpsmen were awarded the Medal of Honor. I can assure you Doctor that every Marine in this division has the highest respect and admiration for your men. We are certainly grateful for their heroic efforts. Doctor, there will always be a tight bond between the Marines and the Navy Medical Corps."

"Thank you sir. We are equally proud to serve with your Marines," replied the division surgeon.

Dr. Hering smiled, bundled up and headed back to the unending line of wounded.

Colonel Bowser looked over at General Smith and said, "General, the pilots have been very busy today and the airstrip is working better than I could have imagined. We sent ten planes back to Yonpo loaded with parachutes."

"Thanks, Al. We'll need them for more airdrops when we get to Koto-ri," replied Smith.

Shortly before midnight, Smith returned to his quarters. Bowser and his staff worked late into the night on the breakout plan to Koto-ri.

The fortunes of war were about to change.

17

"RETREAT HELL"

December 5

HAGARU

At 6:00 am, Smith and Murray got up, shaved and ate a quick breakfast. Murray left to meet with his battalion commanders. A few minutes later, a tired looking Colonel Bowser arrived at Smith's quarters where they reviewed the plan for the breakout to Koto-ri.

"Did you get any sleep last night, Al?" asked Smith.

"Not enough," replied Bowser. "But hopefully a little coffee will take care of me."

At exactly 8:00 am, they put on their parkas and went out into the frigid morning air. They walked briskly the short distance to the command tent. As they walked through the door, all the officers stood to greet General Smith.

"Good morning, gentlemen. Please take your seats," greeted Smith. There were several large pots on the two stoves that heated the command tent. The aroma of the freshly brewed coffee however did not mask the pungent smell of the unwashed Marines. The command tent had the unmistakable smell of a high school locker room. The tent was packed with the officers of the 1st Marine Division: Colonels Litzenberg, Murray, Beall, Partridge and Ridge, all battalion commanders and Smith's division staff.

Also in attendance was Brigadier General Thomas Cushman, the Assistant Wing Commander of the 1st Marine Air Wing.

General Smith sat down, while Bowser remained standing to start explaining the breakout plan.

The colonel started in, "Colonel Murray, your 5th Marines will take over the defense of Hagaru at noon today. Colonel Ridge's battalion will be attached to you. At first light tomorrow, you will attack and push the Chinese off East Hill. We can't leave them on the highest spot around Hagaru when we start moving the men out and our perimeter begins to shrink.

"Meanwhile, Colonel Litzenberg's 7th Marines will lead the way out of Hagaru down the road. Colonel Murray, when you have secured East Hill, you will then follow Colonel Litzenberg's 7th.

"All personnel will walk alongside the motor vehicles to provide protection. The only exceptions to this are obviously the drivers, radio operators and casualties incurred along the way."

Litzenberg interrupted, "What happens if a truck breaks down?"

"If the truck is found to be inoperable, it is to be pushed out of the way and destroyed," replied Bowser. "We already know there are a lot of roadblocks along the way. I want the division engineers up front with the heavy equipment and if we encounter an obstacle, bulldoze it out of the way or blow it up," continued Bowser.

"All troops are to take enough rations for two days. Each rifleman will be issued 100 rounds of ammunition; extra ammunition will be carried on the backs of the trucks."

Murray asked, "What do we do about casualties?"

Bowser replied, "Any casualty will be treated, placed in his sleeping bag and carried on a truck. We will do our best to evacuate all serious cases by helicopter. All supplies and equipment that we can't take with us will be destroyed. Either blow it up or burn it, but we are leaving nothing for the Chinese."

Lieutenant Colonel Youngdale who was now in command of the 11th Marines, the artillery regiment of the 1st Marine Division, stood up and said, "Sir, we have a lot of artillery ammo that we can't take with us. There is just no room for it in the trucks. My recommendation is that instead of destroying it, we shell the hell out of the hills on the road to Koto-ri. I'm sure the Chinese will move back into the hills after dark and then we'll let them have it."

"Okay, let's give it to them and give it to them good," said Bowser cracking a faint smile. Bowser was an old artillery man and loved to see the big guns in action. Bowser continued, "Gentleman, the intelligence gained from POWs is that there are eight Chinese divisions operating in our area. In addition, there's some pretty strong evidence that at least two more divisions are heading our way. And one other thing, we've got about 10,000 men on this breakout to Koto-ri. We'll also have over 1,000 vehicles. We don't have much time; let's start lining them up. Gentlemen, General Cushman would like to say a few words."

Cushman stood, walked to the front of the room. "Thank you Colonel Bowser. Tomorrow at 9:00 am, Marine Air will be on station to furnish close air support for your attack on East Hill. Also, we will have an umbrella of twenty-four aircraft to cover the breakout column all the way to Koto-ri. There will be additional search and attack planes that will hit the ridges flanking the road. All close-in air strikes within three miles of the road will be guided by your forward air controllers on the ground. I can assure you that this concentration of aircraft covering your attack to Koto-ri will be the greatest that this war has ever seen. Marine Air will fly out of Yonpo and also off of an aircraft carrier; in addition, Navy pilots will fly off four other aircraft carriers. Gentlemen, I can promise you one thing. Tomorrow morning we're going to ruin their fucking day!"

The whole command tent broke out in laughter and cheers.

General Smith stood. "Thank you, General Cushman, Colonel Bowser. Gentlemen, we finally have most of our division together. We are now a much more powerful fighting force. We rested the men for a day and tomorrow we set out on the second leg of our breakout. You and your men have done an exceptional job. Every day I hear about the heroics of the officers and men. I can't tell you how these stories of extraordinary courage and leadership humble me. When they write the history of this battle, it will be about these men, but in a larger sense it will be about what it means to be a U.S. Marine. The Chinese will never defeat us!

"Gentlemen, I can't think of anything more. Good luck and I'll see you in Koto-ri. By the way, Colonel Partridge, if you would remain here, I'd like to talk with you."

Everyone left except the division staff. Smith, Partridge, and Bowser sat back down at the table.

Smith asked, "John, what is the latest on the bridge?"

"Sir, I think the best way to repair that bridge is with treadway bridge sections. I think we could air drop these sections at Koto-ri and from there it would be a three-and-a-half mile drive to the bridge," answered Partridge.

"How many sections do you need?" asked Smith.

"Four sections, sir."

"Has anyone done a test drop to see if this would work?" continued Smith.

"They did drop several of these 2,500 pound sections for a test from a C-119 aircraft but not with good results. The Air Force assures me that they have learned from this and that it can be done," answered the colonel. "It is my recommendation that we airdrop eight of these sections and hopefully four of the sections would be usable."

"All right," replied Smith. "Now, how are you going to get these sections to the bridge site?"

Partridge replied, "Sir, there is an Army treadway bridge company that got stuck down in Koto-ri on the night of November 27 and fortunately they have two Brockway trucks capable of carrying these bridge sections. Each truck can carry four sections. Each section is sixteen feet long. Two sections can be hooked together to form a thirty-two foot bridge span. So with four sections, we have two thirty-two foot spans. One forms the left side of the bridge and the other forms the right side with plywood sections making up the middle of the bridge."

"What do you need the plywood sections for?" asked the general.

"The purpose of the plywood sections is to make the bridge capable of carrying any type of wheeled or track vehicle with different size wheelbases so we can then drive everything from jeeps to tanks over it."

"John, how are you going to put these bridge sections onto those Brockway trucks and how are you going to put them in place once you get to the bridge site?"

"Sir, mounted on the back of the Brockway trucks is a big crane which can lift these heavy treadways on and off the trucks."

Smith leaned back in his chair and lit his pipe while thinking the whole process through.

"What if it doesn't work and all these bridge sections are damaged? What do we do then?" Smith continued his questioning.

"General, I have already arranged to have bridge timbers stockpiled at Koto-ri."

Smith was not a micro-manager; he didn't usually go so deeply into details of tasks he assigned to his men, but in this case he needed to know and understand every last detail. That bridge had to work; it was key to the breakout and Partridge's plan was untested.

Partridge was becoming visibly annoyed by all these questions. His frustration built to the point to where he blurted out, "General, I got you across the Han River didn't I?"

"Yes, you did," said Smith.

"And I got you up the road."

"Yes, John, you did," again allowed the general.

"And I got that airstrip built didn't I?"

"Yes, John."

"Then god dammit, sir, I'll get you across that bridge," said the exhausted engineer.

Smith was startled by his response but soon broke into a smile.

There was enormous pressure on both these men. Partridge had a bridge to repair and General Smith had a division to save. Smith told Partridge to go ahead with his plan. Partridge took his leave and walked out of the command tent, still a little miffed to have his competence questioned.

Al Bowser looked at Smith and said, "Colonel Partridge seems pretty feisty this morning, General." Smith shook his head in agreement and just smiled.

Yet another visitor arrived that morning. This one was unexpected. Air Force Major General William Tunner, Commander of the Far East Combat Cargo Command walked in. Tunner was greatly admired by the Marines on the ground for the extraordinary job he had done in resupplying them. He had flown the hump from Burma to China in World War II and in 1948, had also been in charge of the Berlin Airlift. He was warmly greeted by General Smith.

"Bill, what are you doing here?" asked a surprised Smith.

"O.P., I flew here to talk to you about a couple of things."

"Bill," replied Smith, "first I want to offer you my sincerest thanks for the outstanding job you and your men have done resupplying my Marines with your planes. I don't know just how many tons of supplies and ammo you air dropped by now, but God only knows where we'd be without your help. What's up?"

"O.P., as soon as we evacuate all the casualties, I think we could fly planes in here as fast as the airstrip at Hagaru could handle them. We could fly out your whole division."

Smith was stunned by the proposal. He obviously liked and respected General Tunner but this offer made no sense to Smith.

General Smith quickly replied. "Bill, we've flown in over 500 replacements and I don't see any good reason to fly them out. No able-bodied Marine will fly out of here. We're fighting our way out and we're taking all our equipment with us!"

"I understand, General," said Tunner, "but I had to make the offer."

"Thanks again, Bill, for all the help and have a safe flight back," finished General Smith.

The generals shook hands and General Tunner left for the airstrip.

Smith turned to Bowser and said, "I don't think the general understood a couple of things Al, and I didn't want to embarrass him. If we started evacuating our troops by air, the perimeter would continue to shrink till the only thing left would be the airstrip. And the few units that were left wouldn't have the strength to defend themselves against the Chinese closing in on them. Al, that just wouldn't have worked and we sure weren't going to abandon our artillery, tanks and vehicles."

Later, a major who was in charge of the Post Exchange section (PX) entered the command tent and requested to speak to General Smith.

"Sir," said the major, "as part of the withdrawal order from Hagaru, I've been ordered to destroy all the goods in the PX. Sir, this PX was set up originally on the assumption that Hagaru would be established as a base. There is about $13,000 worth of stock in there. And before I destroy it, I wanted to check with you. My orders are that the Chinese get nothing."

General Smith asked, "What's in there, Major?"

"Well sir, we've got lots of candy, stuff like Tootsie Rolls. There is also cigarettes, soap, toothpaste and other things like that."

"Major, pass the word to all the troops and make sure everyone gets the word; they can have whatever they want," replied the general. "Tomorrow morning you can destroy anything that is left."

The major smiled and said, "Yes sir, will do."

The things that the Marines grabbed first were the Tootsie Rolls and they stuffed their pockets and packs full of them. Tootsie Rolls had been in military rations since World War I.

The sugar in the candy provided quick and needed energy and wouldn't cause any intestinal disorders. They didn't melt in the heat and if kept close to the body, they wouldn't freeze either. On more than one occasion they would be used to plug bullet holes in leaking radiators. In the days to come the road south from Hagaru would be littered with Tootsie Roll wrappers. Many men also gave the surplus PX items to the Koreans who still lived in Hagaru and who had shown friendship and courtesy to the Marines.

Early in the afternoon several reporters arrived in Hagaru. With the opening of the airstrip, they were given the opportunity to visit the Marine base.

Among them were Keyes Beech of the *Chicago Daily News*, Marguerite Higgins of the *New York Herald Tribune* and others representing various American newspapers. There were also two French correspondents and a British correspondent. These reporters were jeeped to a large heated tent adjacent to Smith's command tent.

Smith had never been a publicity hound and didn't particularly enjoy talking to the press. Colonel Bowser, however, convinced Smith to talk to them.

"The world should know about the Marines of the 1ˢᵗ Marine Division," said Bowser. "They have been in hell for weeks and we should make them and their families proud of what they have been doing."

"Okay," said Smith. "You made your case."

Smith and Bowser walked to the tent where he greeted all the reporters. He was surprised to see Marguerite "Maggie" Higgins among the group. Smith briefly explained the perseverance and tenacity of his troops and the breakout plan to Koto-ri and then he answered a few questions.

One of the reporters asked General Smith, "What do you think of the Royal Marines?"

To which he replied in his usual understated manner, "Well, they are Marines, aren't they! The last time we fought together was in the Boxer Rebellion in 1900. They're a good outfit and we're glad to have them with us."

The reporter responded, "These Brits look pretty sharp in their berets."

Smith said, "Yes, they are a fine looking outfit. Like our men, these Marines are here to fight."

During the past several days, the press had been filing reports in the American newspapers that the Marines were trapped and for the first time in history, they were starting to retreat.

Following up on these reports, the British reporter asked, "General is this a retreat?"

Smith did not like this word "retreat" especially as it applied to his division. He'd heard about the press reports and they disturbed him.

He carefully replied, "Look, when a unit is surrounded, there can be no retreat. A retreat implies that there is no enemy to the rear. But we're surrounded. We aren't retreating, we're attacking in another direction."

The next day newspapers across America printed some of the most famous words in Marine Corps history—"Retreat hell. We are just attacking in another direction." It would become the line forever associated with the general.

After the press conference was over, Maggie Higgins proudly announced to the assembled crowd her intentions to march out of Hagaru with the Marines. She was a famous reporter and covered the Marines in both the Pusan perimeter and the Inchon-Seoul battles. Even her winter combat clothes couldn't conceal her beauty. When she arrived on the scene, she turned a lot of heads. Many of the male correspondents didn't like her. Some of the generals had a way of confiding in Maggie Higgins much to the chagrin of her competitors. General MacArthur was no exception. General Walton Walker, who commanded the 8th Army at the beginning of the Korean War, ordered her out of Korea because she was a woman. MacArthur overruled him; Maggie Higgins would stay.

"Miss Higgins, you will not walk out with the Marines," General Smith told her. "It's not safe here and it sure isn't safe on that road. You will fly out of Hagaru by nightfall. Understood?"

She raised her voice and said, "I understand, General Smith but it's just not fair."

Higgins was furious but as ordered, she flew out later that day. Smith did not want a woman injured, killed or worse, captured on his watch. She just didn't belong here and he could not guarantee her safety. His plate was full enough without having to worry about the distraction of an attractive woman on a journey he knew would be straight from hell.

18

EAST HILL

December 6

GENERAL SUNG

There were eight Chinese divisions surrounding the withdrawal route to Koto-ri. General Sung, as a result of his huge losses, found it necessary to order his four reserve divisions to move immediately south to join in the Chosin battle. Fortunately for the Marines, Sung had greatly overestimated the ability of his soldiers at Chosin to destroy the 1st Marine Division.

Only two of the four divisions arrived in time to battle the Marines. Had all four divisions arrived at the same time, the final outcome might have been different. Sung's troops had been slowed by air attacks and some had gotten lost in a snow storm. The advance elements of the Chinese troops had made it to East Hill and tried desperately to hold it. The cold had killed off many of the coolies who carried the supplies. General Sung's widespread dispersal of his troops and supply lines prevented the Chinese from being able to strike the Marines in full force. This would prove to be a critical error for General Sung.

In anticipation of the Marines' attack to the south, the Chinese had already set up nine roadblocks in the eleven miles between Hagaru and Koto-ri. They were primed and charged to accomplish the task of annihilating the 1st Marine Division.

HAGARU

Unit tasks had already been assigned. Puller's 1st Marines would continue to hold Koto-ri, the 5th Marines would defend Hagaru and clear East Hill while the 7th Marines would move down the eleven miles of road to Koto-ri. At 5:00 am, a heavy fog covered the base at Hagaru as Ray Davis' 1st Battalion 7th Marines led the way down the icy road.

At 7:00 am, Ray Murray's 5th Marines attacked East Hill. The attack was preceded by Marines pounding the hill with mortars. A half hour later, the Marine and Navy aircraft arrived on station flying through the fog, hitting East Hill with machine guns, rockets and bombs. The devastating air assault lasted a full ninety minutes which would prove to be the largest air operation during the Chosin Reservoir campaign.

General Smith woke early and was already having breakfast under the ever present picture of Joseph Stalin when Lieutenant Colonel Partridge reported in to discuss the blown bridge.

"Good morning, General," greeted Partridge.

"John, have a seat," replied Smith. "How is the weather this morning?"

"It's bitter cold. There is a heavy fog overhanging the base," answered the Colonel. "As soon as the fog lifts, I will take off to fly over the bridge site to confirm that there are no changes. I want to be sure things remain the same as the reconnaissance report on December 4. A lot could change in two days. After making the recon I plan on landing at Koto-ri. When you get there, General, I'll fill you in on what I've learned and I'll also meet with Colonel Puller to schedule the bridge drop."

"That blown bridge is the biggest obstacle to getting us out of here."

"I understand sir," said Partridge as he left General Smith.

Partridge was driven down to the airstrip he had built. He had mountains to move and little time to move them.

General Smith, after finishing his breakfast, put on his helmet and heavy parka and walked out into the frigid morning air to the command tent where he met with Al Bowser.

"Sir," said Bowser, "the 7th Marines are already on their way to Koto-ri and the 5th Marines are attacking East Hill. The engineers are in the process of destroying anything that we are not taking with us."

"Al, when Litzenberg lets me know they've got a good start, then we will fly the command group to Koto-ri to relocate and set up the division command post."

Smith remained in the command tent to closely monitor the progress of the 5th and 7th Marines.

Shortly after 9:00 am, Lieutenant Colonel Partridge flew out of Hagaru on a small, unheated observation plane. As the plane climbed in the air, Partridge could see bombs from the Marine fighter planes blasting away at East Hill. He could also see the long line of trucks stretching almost a mile down the road. As the plane flew further south, he saw several roadblocks. After passing Koto-ri, he also noticed a large hole in the road, something else his engineers would have to deal with. Partridge could now see the bridge site from the air. He flew over and around the bridge for twenty-five minutes. Seemingly nothing had changed; the gap remained about sixteen feet. He then flew from the bridge on to Chinhung-ni which he knew would be the next stop on the breakout after Koto-ri.

KOTO-RI

Partridge's aircraft turned around and landed at Koto-ri, arriving there at quarter till ten. He had observed much but couldn't make any notes since his hands were almost frozen. Every time he tried to hold his pen, he found his fingers would not cooperate.

Upon landing, Lieutenant Colonel Partridge was driven to Colonel Puller's command post. The heat from the tent was most welcome after the unheated plane and jeep ride. Puller sat in the tent smoking his stubby pipe.

"Good morning, Colonel Puller," greeted Partridge.

"How was the flight down here?" asked Puller.

"Colonel, I've never been so goddamn cold in all my life," answered Partridge. "Sir, I understand you have been updated on the plans to repair the bridge at the pass. I just want to be sure we have enough sandbags and timbers in case the airdrop of the treadway bridge sections doesn't work."

"I can assure you, John, I have stockpiled enough of them to build you all the goddamn bridges you want," replied Puller. "One other thing, John. I assume there will be a lot of casualties from the 5th and 7th Marines as they fight their way here. I'd like your engineers to start lengthening the airstrip here at Koto-ri. As it is now, we can only land helicopters and light observation planes. We're going to need those transport planes for evacuations. To accomplish this, how many more feet will you have to add to the runaway?"

"Well sir, I think another 400 feet will do it," replied Partridge.

"Ok, get your engineers on it right away," finished Puller.

Partridge's to do list kept getting longer. He wondered if there would be enough hours in the day and enough engineers to get the job done.

In addition to his duties as the commanding officer of the 1st Marine Regiment, Puller also had to assume duties normally reserved for an assistant division commander. When the 5th and 7th Marines arrived, the base at Koto-ri would triple in size. He had to ready the base for the arrival of an additional 10,000 troops and 1,000 vehicles. These men would be cold, hungry, exhausted and many would be wounded.

Shortly before 2:00 pm, General Smith received a positive call from Colonel Litzenberg that he had already moved two miles down the road.

Smith said to Bowser, "Okay, Al, let's move the command group to Koto-ri."

A few minutes later, Smith flew to Koto-ri by helicopter. Bowser followed right behind him. It was a short ten minute flight. A lot faster than the 7th Marines were able to move, thought Smith. Upon landing, they were taken to Colonel Puller's command post.

The Savior

When they entered the CP, Colonel Puller greeted them, "General Smith, Al, welcome to Koto-ri."

"Lewie, it's good to see you. Will you update us on the plans to bring in Murray and Litzenberg?" asked Smith.

"Yes, sir," replied Puller. "The warming tents are set up and we have enough galleys to supply all the troops with hot chow. As you can see, we're packed in here pretty tight but we'll be able to handle everybody. I have already instructed Colonel Partridge to lengthen the airstrip so we can get the transport planes in to evacuate casualties. If this airdrop of the bridge sections doesn't work we have plenty of building timbers on hand. The medical facilities are ready to accept any additional casualties. I've got about 2,500 Marines here plus about 1,500 soldiers. My men are in good shape and we're holding a heavily fortified perimeter. You know, General, there's one good thing about being surrounded."

Perplexed, Smith looked at Puller. "What's that?" asked Smith.

"Hell, sir, now we can shoot 'em in every direction."

Smith smiled. Puller was always ready with a colorful comment. Smith was encouraged to hear his regimental commander ready to take on the entire Chinese army. He expected nothing less of Puller.

"Sir, your CP is located right in the middle of the perimeter. This large tent and some other gear had been sent up here by X Corps before the battle started. This was going to be X Corps' advance CP. I decided that it would make a fine command post for you so I commandeered the tents. This will also be your sleeping quarters."

"Thanks, Lewie. I appreciate that. As always, you are on top of things."

Smith and Bowser left Puller and were jeeped to their new division headquarters to start planning for the move to Chinhung-ni, ten miles south of Koto-ri.

When Smith walked into the command tent, he said to a captain standing nearby, "Find Lieutenant Colonel Partridge and have him come up here right away and please get our maps set up in here."

"Aye aye, sir," was the reply.

Ten minutes later, Partridge reported to the command tent.

"John, what's the status on the bridge and the air drop of the bridge sections?" asked Smith.

"Sir, the gap is still sixteen feet and as I told you earlier, the Air Force has done a trial drop of a bridge section but it didn't work out. After that, however, special parachutes have been flown into Yonpo Air Base from Japan with a crew of expert Army parachute riggers. They have assured me that this drop will work successfully," stated Partridge.

"John, when will the air drop take place?" asked General Smith.

"First thing in the morning, sir."

"Have you cleared this with Colonel Puller?" Smith inquired.

"No, sir."

"You need to do that right away," continued the general. "After all, Colonel Puller is the base commander and an air drop with these heavy sections could result in casualties. These men have faced enough without the sky falling on them. I think it would make more sense to schedule this drop later in the day. This will give Colonel Puller time to alert his men that the drop is going to take place and they should be on the lookout for it."

The engineering officer gave a rare smile and left to clear everything with Colonel Puller. Partridge didn't want to delay the drop too long because there was much work to do once the sections arrived. He was sure Puller could get his men alerted quickly.

HAGARU

As the 7th Marines slowly fought their way down the road, the 5th Marines were still fighting to secure East Hill.

Although all casualties had been airlifted out the day before, wounded from the day's fighting would require last minute evacuations. Sixty new casualties were flown out before the air field was shut down at 5:00

pm. No more planes would fly from this life saving strip that Partridge and his men had created and that many pilots had risked their lives to fly into. In total, 4,312 casualties including 3,150 Marines, 1,137 Army soldiers, and twenty-five Royal Marines were flown from this airfield before it was abandoned to the Chinese. The airfield was a nearly impossible engineering feat that had saved the lives of many Marines. General Smith's planning and preparation had served his men well.

The battle for East Hill would take twenty-two hours. In spite of enormous casualties, the Chinese were not about to give up control of East Hill. They attacked again and again until midnight when they pulled back to reorganize. At two in the morning, they attacked with a vengeance. The next three hours witnessed the most violent and destructive battle of the entire Chosin Reservoir campaign.

Under Lieutenant Colonel Ray Murray's command, Marine tanks, artillery, mortars, machine guns and riflemen did their deadly work as they cut down the unending stream of the white uniformed Chinese.

Murray watched the waves of Chinese soldiers falling and dying on the hill.

Why did General Sung so recklessly order his men into the jaws of certain death, he wondered. Murray questioned, did the leader really think he could hold East Hill? Did it matter how many of his men he sacrificed? Was he afraid of failure?

Whatever the reason, Sung had made a terrible and costly decision because the Chinese never had a chance against the overwhelming firepower unleashed by the Marines. The Chinese ran out of soldiers before the Marines ran out of bullets. The early morning light would reveal a scene of unbelievable slaughter. As many as 800 dead Chinese soldiers littered East Hill. Blood stained snow surrounded each body in the surreal landscape. The Chinese had indeed suffered a major defeat and the battle to take East Hill was finally over.

19

A GREAT LOSS

December 7

KOTO-RI

At 5:00 am, the leading elements of the 7th Marines began arriving at Koto-ri. As they neared the camp their exhausted shoulders began to rise under their burdens. They started to look more like the proud Marines that they were. Through the frozen morning air, the sound of their cadence was music to the Marines defending Koto-ri. The defenders began cheering as more and more of the 7th Marines came through the perimeter. These men had battled through the night and the bitter cold. They had lived to fight another day.

This morning like all the others started early for General Smith. He went to the perimeter to see the men of the 7th coming into Koto-ri. It was hard to believe that these were the men who looked so young to him when they landed at Inchon in September. They had been the last regiment to join the fight and when he saw them they looked like school boys. Now they all looked old. War and the unrelenting cold had stolen their youth. In two-and-a-half months, the fresh faces had disappeared. The sights they had seen, the hardships they had endured and the constant danger they had faced had aged them prematurely.

Smith was already in the command tent puffing on his pipe when Colonel Litzenberg reported in. Twenty-four hours of constant marching and fighting in below zero temperatures showed on Colonel Litzenberg's unshaven face. Taking his cigarette out of his mouth, he smiled and said, "Good morning, General; 7th Marines reporting in sir."

"It's good to see you, Litz. How did it go out there last night?" asked General Smith.

"Sir, we had a rough night. About 2:00 am, the column was halted while we were clearing the road ahead. The regimental command group started taking heavy fire from some huts on the left side of the road. We returned fire and the huts started burning, lighting the place up like the Fourth of July. Two of my officers were killed; my executive officer took a bullet through the ankle and sank to the ground. My regimental chaplain climbed into an ambulance to console a badly wounded Marine when machine gun bullets shattered his jaw and killed another Marine next to him. We beat off the attack and moved on. I've appointed Lieutenant Colonel Davis to take over as my XO."

Taking a long draw on his cigarette and slowly letting the smoke out, Litzenberg said, "General, I've got some more bad news."

"What's that?" asked Smith.

"Sir, the Commanding Officer of my 3rd Battalion is missing. Lieutenant Colonel Harris."

"What? General Harris' son? The Marine who had been a POW in World War II? Are you sure?" Smith was stunned. He said a silent prayer that this was one notification he would not have to make.

"Yes sir. I got a call on the radio at 5:30 am that Harris had disappeared. So I sent a platoon up the road with orders to search a ravine where Harris was last seen. They called his name over and over with no answer. They went up that ravine a lot further than they should have but still no sign of Harris. I've ordered my men to check the medical facilities and graves registration, but nothing yet."

Smith lowered his head and thought to himself all that young Harris suffered in World War II and all that his father, General Harris, has done for us and now this. It's just so wrong.

Then he looked at Litzenberg and said, "Litz, I have to report this to General Harris but I don't want to do it until we are absolutely sure his son is missing. Come back to me at 5:00 pm and hopefully he will have turned up."

"Yes, sir," replied Litzenberg.

At 9:00 am, Lieutenant Colonel Partridge entered Smith's command post.

"Are you ready?" asked the general.

"Sir, the bridge sections are scheduled to be dropped at 9:30 am."

"So early? Has Colonel Puller been informed of this?" questioned General Smith.

"Yes, sir, he has and he has alerted all his men to be on the lookout for the air drop."

"Alright John, get back to me when the drop has been completed and let me know how it went," directed General Smith.

Half an hour later, Smith put on his winter gear and walked out into the cold. The big C-119s were right on time. He could see the parachutes open as three of the treadway sections were being dropped on the eastern side of the Koto-ri perimeter. The other five sections were dropped to the west. Smith could only hope that the air drop was successful. They were so big he couldn't imagine how they were going to land without breaking. He walked back inside the command tent waiting to hear from Lieutenant Colonel Partridge. He found himself hoping both for a successful airdrop and finding Bill Harris alive and well.

An elated Partridge returned shortly and reported to Smith.

"Sir, the drop worked!" exclaimed Partridge with jubilance rarely seen in the serious engineer. "Six of the sections landed unharmed; one got banged up pretty bad and we can't use it. The wind carried one beyond the perimeter into the hands of the Chinese. They sure can't do anything with it! We loaded the six unharmed sections onto the Brockway trucks. We also recovered the plywood center sections."

"Great news, John! How is the airstrip coming?" asked Smith.

"My engineers have worked through the night and we now have the Koto-ri runway long enough to handle the big planes. Fortunately, this was not the nightmare job that Hagaru was. The wounded are being evacuated as we speak."

"John, as usual your engineers have done a magnificent job. Please tell them well done," replied Smith.

"Thank you, General. I'll pass it along," responded Partridge as he walked from the command tent with an unusual bounce in his step.

Colonel Bowser, on hearing Colonel Partridge's report to General Smith, turned to the general and said, "Now that the Air Force has flown in the bridge sections, they are going to start resupplying us. Since we now have almost the entire 1st Marine Division together, I have requested airdrops this afternoon of rations, fuel, ammo, hand grenades and medical supplies. All together they will drop about 500 tons of materials."

"Is that going to be enough?" questioned Smith.

"I think it will sir," replied Bowser. "Given that some supplies will be damaged when they hit the ground and some might drift into enemy hands, I still feel that this amount will be sufficient to handle our needs."

"Al, how is the breakout plan coming?" asked Smith.

"I'm almost finished with it and I've scheduled a meeting with all regimental and battalion commanders and division staff for 7:00 pm."

Smith stood, paced back and forth while puffing his pipe. He stopped and looked at Bowser. "The road between Chinhung-ni and Koto-ri is the most dangerous and challenging section of this whole road and it's also the narrowest part. I've driven over this ground twice. Al, if something happens to one of these tanks, a breakdown or being hit by enemy fire making it inoperable, it will block the whole road. We want to be sure that the tanks are the last thing to leave Koto-ri."

"I understand sir, and will be sure it is clear in the breakout plan," said Bowser.

In the middle of the day, a surprise visitor showed up at Colonel Puller's CP. It was Maggie Higgins. She announced to the first Marine she saw, "I'd like to see Colonel Puller."

Surprised by the presence of a woman, the sergeant went to the back of the command tent and said, "Colonel, there is a woman here who wants to see you."

"What? A woman? Who is she?" asked a somewhat confused Puller.

"I didn't ask sir, but she's a looker," answered the sergeant.

"All right send her back," grumbled Puller.

Maggie was escorted by the clerk to Colonel Puller at the back of the tent. The sergeant reluctantly left them alone.

"Colonel Puller, I'm Marguerite Higgins from the *New York Herald Tribune*."

Puller, of course, had heard of Maggie Higgins but he had never met her.

He quickly replied in his gruff, impatient tone, "I know who you are. But what are you doing here? We are in the middle of a war in case you haven't noticed."

"Colonel, I'd like to ask you some questions."

Puller reluctantly deferred to the woman reporter. He motioned to a chair and said, "Have a seat."

As a matter of courtesy, she asked a series of superficial questions which Puller answered. Maggie then offered an observation followed by yet another question.

"Colonel, as you may know, I was at Hagaru when the 5th and 7th Marines came in from Yudam-ni. They were a haggard group of men. But before I came into your command tent just now, I was driven up the road where I watched a column of the 7th Marines marching into Koto-ri. There is an unmistakable difference in the men arriving at Koto-ri from what I saw coming in to Hagaru. These men had more spirit and purpose to their step. Colonel, how do you account for this?"

"Well, Miss Higgins, it's pretty simple. These Marines are savvy individuals and I guess they figured that if they made it this far, then by God they are going to make it the rest of the way."

Puller pulled out his pipe and lit it and in the most diplomatic tone he could manage, he continued, "Miss Higgins, this is a very dangerous place. The Chinese know it's their last chance to stop us and the fighting is going to be fierce. You're going to have to leave here."

She protested, "Colonel, this is the biggest story of the Korean War and I don't want to miss it."

"I'm sorry Miss Higgins. It is not safe here and you will have to leave," replied Puller somewhat impatiently. He was not used to being questioned.

Just to be sure that his directive was carried out, Puller assigned a young lieutenant to make sure she flew out before nightfall.

Maggie Higgins stormed out of the command tent trailed by the lieutenant. Puller thought the young officer might wish he was going head to head with the Chinese before this assignment was over. Higgins was furious but she wouldn't give up so easily; she never did. To be rebuffed twice just wasn't going to work for Maggie Higgins.

Shortly before dark, the worn out lieutenant reported back to Puller.

"Is that woman on the plane out?" asked Puller.

"Yes sir, Colonel. She was none too happy but she is on the plane. I don't think that I'd like to be that pilot though; she was pretty mad."

"Good work, Lieutenant," said Puller. "This is no damn place for a woman."

While Puller was dealing with Maggie Higgins, Smith and Bowser worked on the breakout plan. They both knew once they arrived at Chinhung-ni, the division would be safe. It was ten miles from Koto-ri to Chinhung-ni and it would be the toughest ten miles of the breakout. Both the topography of the road and the fact that the Chinese knew this was their last chance to succeed added greatly to the danger of this final leg of the journey.

Late in the day, Dr. Hering arrived at Smith's new division command post.

"General Smith, thanks to the expanded airstrip, we flew 200 casualties out today. Hopefully we can get the rest out tomorrow," reported the doctor.

"How many more casualties are left to fly out?" asked Smith.

"Sir, as you know, the 5th Marines have yet to arrive and I know there will be more wounded. But what I don't know is how many. With these sub-zero temperatures, we're also going to have a lot more frostbite cases. It's going to take time to screen these men as they come in."

"How is your medical team holding up, Doctor?" asked the General.

"Sir, they're doing a remarkable job but like everyone else around here, they are just plain worn out. Too many wounded and the bitter cold is taking a terrible toll."

"Doctor, your team has done a terrific job and we're almost there. We just have to hang on a little longer," added Smith.

"Thank you sir. Well, General, I'd better get back to the medical tent," and the doctor took his leave.

Fifteen minutes later, the tent door opened again and Colonel Litzenberg entered Smith's CP along with a blast of frozen air.

Smith looked up and could tell by the expression on Litzenberg's face that he was not bringing good news.

"Sir," began Litzenberg, "we have checked everywhere and there is absolutely no sign of Colonel Harris."

Smith knew he had to make the call, but it was the last thing he wanted to do.

He stood up and walked over to the radio and told the clerk to call the 1st Marine Air Wing and get General Harris on the line.

A few minutes later the clerk interrupted, "General Smith, General Harris is on the radio."

Smith walked over and reluctantly picked up the handset.

"Field this is O.P. Smith."

"O.P., how are you doing up there?" asked Harris.

"We are holding on, Field, but I didn't call about that. Field, I've got some bad news. We can't find Bill."

There was a long pause on the other end. Finally Harris managed to say, "What happened?"

Smith carefully explained where Bill Harris was last seen and went into great detail concerning the circumstances of his disappearance and the thorough search for General Harris' son.

Finally, General Smith said, "Field, we haven't given up hope. We're still looking."

General Harris could barely get the words out. "Thanks O.P. I know this had to be a tough call for you."

"Not nearly as tough as it is for you, Field," responded Smith. "I am so sorry."

At 7:00 pm, the command tent was packed with regimental and battalion commanders and the division staff as Colonel Bowser presented the details of the breakout plan.

"Colonel Litzenberg, your 7th Marines will lead the way. You will move out at 8:00 am tomorrow. After this meeting you'll get together with Lieutenant Colonel Partridge to coordinate his engineers' movements down to the bridge and review his plans for the bridge repair. You will be responsible for the engineer's equipment. You have seen the trucks carrying the bridge parts and you know how crucial these are to our success.

"Colonel Murray, your 5th Marines will follow the 7th.

"Colonel Puller, your 2nd and 3rd Battalions of the 1st Marines will be the last out of Koto-ri. Your 1st Battalion down at Chinhung-ni is the only fresh battalion in the entire division. Hill 1081 is the highest point between Chinhung-ni and Koto-ri and intelligence indicates it's heavily reinforced by the Chinese. It is six miles from Chinhung-ni to Hill 1081. Your 1st Battalion will move north at 2:00 am and be ready to attack the hill at daylight. It is critical that this hill be secured before the 7th Marines get there.

"Tanks will go last because if one of those tanks becomes disabled, it will block the road with no way to get around it.

"Colonel Puller, you will provide security for the tanks." Puller nodded in the affirmative.

"Colonel Partridge, after the last tank and last Marine have crossed that bridge, I want your engineers to blow it to pieces.

"Everyone has their assignments. Any questions? Now General Smith has something to say."

"Gentlemen," started Smith, "we all know that this will be the most challenging terrain on our breakout. We fought our way out of Yudam-ni and Hagaru and tomorrow we will fight our way out of Koto-ri. This is going to be the ten mile trip you will remember for the rest of your lives.

"The Chinese certainly know this will be their last chance to get us and I expect the fighting to be fierce. Intelligence reports that additional Chinese units are massing to the east, west and south of us. That leaves us pretty well surrounded.

"Gentlemen, we've come a long, hard way and by the grace of God we'll make it."

Smith slowly looked around the room at his commanders with heartfelt admiration and said, "Good luck, gentlemen and Godspeed."

Later that evening, Partridge attended a meeting conducted by Litzenberg. Partridge explained to the officers present the plans for the bridge repair and the role the Brockway trucks would play. No one was thrilled with the prospect of having to shepherd these slow cumbersome vehicles through terrain that was going to be crawling with Chinese soldiers. But everyone knew that without the bridge repairs over a thousand vehicles wouldn't be going anywhere. So protecting the Brockways was the first order of business. The meeting lasted forty-five minutes and Partridge left, confident that his trucks were in good hands.

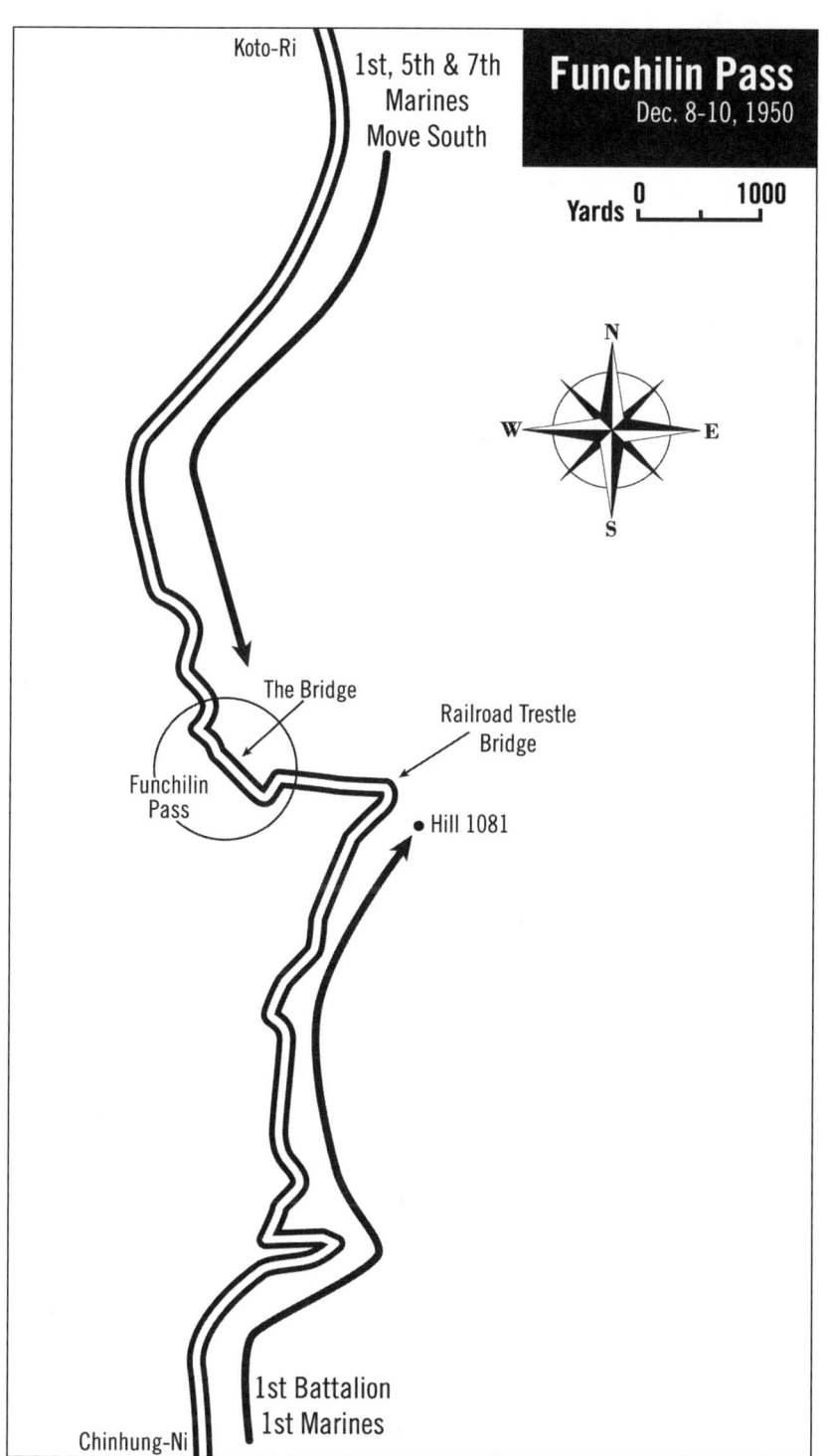

Koto-Ri

1st, 5th & 7th
Marines
Move South

Funchilin Pass
Dec. 8-10, 1950

Yards 0   1000

N
W   E
S

The Bridge

Railroad Trestle
Bridge

Funchilin
Pass

• Hill 1081

1st Battalion
1st Marines

Chinhung-Ni

Back at the command tent, Smith and Bowser continued working on last minute details. They both heard singing and stopped talking. The music was crisp and clear through the frigid air. Smith paused and asked, "Al, who is singing?"

Bowser went outside to check and came back to report, "Sir, those are the truck drivers in the tent next to us."

It was an emotional moment for Smith and he said, "Al, they're singing the Marine Corps Hymn."

"General, in view of all that these men have gone through, they haven't lost their spirit. I think our troubles are almost over. With Marines like this, we've got it made. The Chinese don't have a chance," said Bowser.

Smith grinned and replied, "Bowser, those Chinese never did stand a chance."

Shortly before midnight, the last of the 5th Marines entered the safety of the Koto-ri perimeter. Like the men of the 7th who had arrived fourteen hours earlier, these exhausted Marines were provided hot food and warm tents and a chance for a few hours of rest. The Marines were not out of the woods yet but now they had the momentum. The division was finally back together!

20

HILL 1081

December 8

CHINHUNG-NI

While the rest of the amassed forces at Koto-ri prepared to head south, at 2:00 am, Puller's 1st Battalion in Chinhung-ni started its six mile march north to Hill 1081. They only took three vehicles: a radio jeep and two ambulances. This battalion was well rested and eager to fight having participated little so far in the battle for Chosin Reservoir.

Their mission was to dislodge the Chinese who were holding Hill 1081 on the southern side of the Funchilin Pass. From this position the enemy could prevent the bridge repair and block the troops on the road from reaching Chinhung-ni.

The Marines started out in a snowstorm which reduced visibility to almost zero in the pre-dawn hours. Snow clung to the men's parkas making them almost invisible. Howling winds muffled the sound of their boots as they hiked up the icy road in the negative fourteen degree temperature.

The Chinese on Hill 1081 who were hunkered down against the storm in their foxholes didn't have a clue that the Marines were closing in on them from the south. They were expecting the Marines to come from Koto-ri in the north.

The climb from Chinhung-ni to the base of Hill 1081 was about 1,200 feet of elevation. However, to reach the top of the hill, the Marines would have a brutal, exhausting climb of another 1,000 feet.

At 8:00 am, with the snow still blowing, the Marines reached the base of the hill and started their ascent.

KOTO-RI

After a peaceful night at Koto-ri, the 7th Marines jumped off at 8:00 am as scheduled. They had a day to rest and warm themselves, but then needed to regroup and prepare for yet another hellish day. They were dismayed to see the falling snow as they left their tents early that morning. While the snow covered everything, it made the desolate camp even more miserable. Not only did the snowstorm make their movement difficult, more important it also prevented much needed Marine air support. The men knew that air support was a crucial asset against the Chinese.

As Litzenberg's three battalions moved down the road, they soon became snow covered. Their eyebrows and facial hair iced over and their feet and fingers were freezing. To distract them from the misery of the cold however, there were plenty of Chinese soldiers. The fighting was heavy and the going was slow. The constant crackling of rifle and machine gun fire along with the less frequent boom of mortars were all that could be heard over the howling wind. One battalion would clear the ridges to the left of the road and another battalion would clear the ridges to the right. The third battalion would move down the road followed by the regimental convoy.

At midmorning, General Smith was driven to the medical facility to confer again with Dr. Hering. When he walked into the medical tent, he saw the doctor working on a wounded Marine. Smith waited.

A few minutes later, the doctor finished, took off his surgical gloves, tossed them on the table and walked over to General Smith.

"Good morning, General. It was another long night. There were a lot of new casualties from the Marines fighting their way down from Hagaru."

"How many?" asked Smith.

"The 5th and 7th Marines brought in about 500 casualties sir, but as you know, we flew 200 of the worst cases out yesterday."

"Doctor, with this storm, I doubt if we'll get anybody out today."

Dr. Hering added, "Graves registration now has 117 dead and three still missing including Lieutenant Colonel Harris."

"Doctor, there's no way we can fly out the dead. All available space on the planes and the trucks is needed for the living. Much as we all hate the idea, we are going to have to bury our dead here. Graves registration has already been instructed to transport these men to the burial site. We'll hold a short service there today."

The doctor could see the strain on the general's face. Hering saw just how difficult this was for O.P. Smith. Marines don't leave their men behind.

General Smith returned to his CP. Colonel Bowser informed the general that the 7th Marines were meeting stiff resistance and progress was very slow. The news was not unexpected, but Smith knew that he could rely on Litzenberg's men to get to the bridge. He had to count on them. They were the way out.

An aide came in and delivered a message to Bowser. The colonel looked grim as he announced, "General," said Bowser, "everything is ready at the burial site and the funeral service is scheduled for 10:30 am."

"Al," started General Smith, "I hate burying these men here in foreign soil. They surely deserve better. But we have no other choice."

Hard choices are what Marine commanders had to make.

Half an hour later, Smith was driven to the service. The burial site was a former artillery command post which had been evacuated and now was a large hole in the frozen earth about the size of a small basketball court. Lack of time and the frozen ground prevented preparation for individual graves. In addition to the commanding general, Colonels Litzenberg, Puller and Murray were also present. Each of these leaders took every Marine death personally.

In the swirling snowstorm, the burial detail unloaded the bodies of the 117 Marines, Navy Corpsmen, British commandos and U.S. Army soldiers. Each was wrapped in a sleeping bag or a parachute. Protestant services were held first followed by a Catholic service. Two bulldozer operators then pushed chunks of frozen earth, as reverently as possible, over the mass grave. Smith felt a bitter taste in his mouth. He hated leaving his men behind in this frozen wasteland so far from home.

An artillery aiming stake became the sole marker for the grave site. The location of the grave was surveyed and recorded on a map of the area. Smith knew that if it was ever politically possible, these bodies would be recovered and brought home.

On the way back from the burial service, Smith heard a plane but couldn't see it because of the heavy snow falling. He told his driver to stop.

Suddenly, the plane appeared. Smith watched admiringly as the plane landed safely. It took off a half hour later evacuating nineteen casualties. It would be the only plane able to land that day. Smith heard the plane take off and felt a swell of pride for the courage that surrounded him. Pilots landing in blinding snowstorms, and men holding a hill while riddled with wounds. These were not exceptional feats in this war. He had seen and heard of them on almost a daily basis. Smith offered a silent prayer of thanks for God's help.

Shortly before noon, Murray's 5th Marines followed Litzenberg's 7th Marines out of Koto-ri. Murray's men had only a few hours to recover from their last battle. But they had a hot meal, minimal sleep and now it started again: numbing cold, constant marching. They operated on pure grit.

Back at General Smith's CP, word was received that Colonel Litzenberg's 7th Marines had moved far enough down the road to now start sending the Brockway trucks toward the bridge site. The first truck was loaded with four treadway bridge sections and the second truck carried two sections.

Partridge and his engineering group left Koto-ri with an MP escort. Because of the fighting farther up the road, the Brockway trucks made almost no progress. At 5:00 pm, in the hills of North Korea, it began to get dark. Litzenberg ordered Partridge to turn the Brockway trucks around and bring them back to his CP so their precious cargo could be protected for the night. When they arrived in the vicinity of Litzenberg's CP, they attempted to get the trucks off the road by pulling them into what looked like a rather large parking area. There were already some trucks parked on it. Partridge found that what had appeared to be a parking lot was actually a small pond. The heavily loaded Brockway truck broke through the ice and became stuck. Fortunately the truck carrying the four bridge sections, enough to complete the bridge repair, remained on the road.

Litzenberg ordered Partridge to take the operable truck back to Koto-ri where it would be safer for the night. As the truck lumbered back, Partridge fumed over a wasted day. He had a bridge to repair.

HILL 1081

The unrelenting snowstorm that prevented air cover and evacuations of the wounded was actually a stroke of good luck for the Marines of Puller's 1st Battalion in their attack up the southern slope of Hill 1081. And it was also an important element in helping to surprise and overpower the Chinese strong point on the ridge just below the crest of the hill.

The task of leading the main attack up Hill 1081 was given to Captain Robert Barrow, the commanding officer of A Company, 1st Battalion 1st Marines. In World War II, Barrow served as an advisor to the Communist guerillas in central China. He probably knew more about the Chinese soldier than anyone else in the whole 1st Marine Division.

Earlier in the Korean campaign, Barrow had been awarded the Silver Star for his dramatic stand against the North Koreans on the outskirts of Seoul.

Before the assault on Hill 1081, Barrow told his men, "These are peasant soldiers accustomed to unusual hardships. They can march great distances with ease in the roughest terrain. They can exist on a diet so meager so as to astonish you. They can outwalk us but they sure as hell can't outfight us. All right Marines, let's go!"

A steep slope, loose rock, thin mountain air and bone-numbing cold drained the strength of the bulkily clad Marine riflemen as they pulled and clawed their way up the icy hill. Barrow and his troops climbed up the ridge against no enemy resistance. The ridgeline was so narrow that his men had to walk in single file. With a break in the snowstorm, Barrow was able to see a series of Chinese bunkers on a knob between his company and the crest of the hill.

Barrow's Marines quickly attacked, surprising and killing more than sixty Chinese soldiers with the rest fleeing. He lost ten of his men killed and eleven wounded. The snowfall finally ended and the night turned clear and bitter cold. The first skirmish against the Chinese had alerted them to the Marines presence. There would be no more surprises, just brutal fighting.

At midnight, the Chinese tried to push the Marines off the knob of the hill with a counter attack. Fighting was intense and when the battle was over, Barrow radioed his battalion commander, "We killed them all."

That night the temperature dropped to twenty-five below zero and the wind came roaring out of the north and slammed headlong into Hill 1081. It would be a long, miserable night for the exhausted, frozen Marines. They took shelter in the fox holes previously occupied by the now deceased Chinese.

The crest of Hill 1081 still remained in Chinese hands. When the sun came up the next morning, Puller's 1st Battalion would have to attack the top of the hill and take it. And now there would be no element of surprise working for them. These Marines might not have seen much action previously in the Chosin campaign, but now they were certainly going to earn their pay.

KOTO-RI

Smith and Bowser were still in the command post reviewing the events of the day. Smith was smoking his pipe, the familiar aroma of Sir Walter Raleigh filling the air as he studied the map in front of him.

Bowser said to Smith, "Sir, the 7th Marines didn't make much progress today. Fighting was extremely heavy all day as we expected it would be and we had to bring the Brockway trucks back here to Koto-ri to the safety of our perimeter. The good news is that Puller's 1st Marines made it almost to the top of Hill 1081. I'm confident they will take the crest in the morning."

"You know, Al," said Smith, "that snowstorm was a mixed blessing. It may have helped hide the 1st Battalion on Hill 1081 but it surely slowed the advance of the 7th Marines. It also kept us from evacuating the wounded, and getting air support. Let's just hope that tomorrow brings better weather."

21

THE BRIDGE

December 9

KOTO-RI

The morning started out cold, clear and bright and it gave General Smith a wave of optimism. He turned to Colonel Bowser and said, "Al, I have a good feeling about today. Let's get this show on the road."

Bowser smiled. "Aye aye, sir."

Marine aircraft came on station bombing and strafing the hills above the road and the 7th Marines moved out heading south.

THE BRIDGE

Lieutenant Colonel Partridge assembled his engineers and with the Brockway trucks arrived back at Colonel Litzenberg's command post at 8:00 in the morning. He was told to wait there. Partridge was informed that he should move on down the road as soon as the infantry had taken the high ground at the top of the hill. Litzenberg would then stop traffic and bring the engineers and Brockway trucks to the front to fix the bridge.

At 10:00 am, Litzenberg called Partridge and told him to start moving the bridge sections. Partridge and his team reached the head of the column an hour later. Partridge climbed from his jeep and walked behind a bulldozer as it cleared the snow off the mountain pass.

As they moved along, they came across a huge crater that the Chinese had blown in the road. This was the same hole that Colonel Partridge

The Savior

had detected on his aerial reconnaissance on December 6. The column was going nowhere. The engineers came up with a novel solution: fill it with snow. There was plenty of that to spare. A bulldozer came forward and scraped snow into the hole and packed it down. After forty-five minutes of hard work, the ten-foot deep crater was filled and the column moved forward.

Litzenberg's men had to stand in the bitter cold until the engineers did their work. Hurry up and wait and freeze was the complaint heard up and down the line. The only men who were sheltered in vehicles were the drivers, radiomen and the wounded. Everyone else walked to keep warm and to always be prepared to handle an enemy attack.

When Partridge reached the bridge site, he was stunned at what he saw. The Chinese had blown the bridge again. The gap had widened from sixteen feet to now twenty-nine feet. The treadway bridge sections wouldn't be long enough. All that work and planning appeared to be for nothing. But Partridge didn't give up easily.

Fortunately there was a ledge about eight feet below the blown end of the bridge. One of the engineers had found a stockpile of railroad ties next to the bridge site. Chinese prisoners were ordered to carry the timbers to the ledge. Partridge used the timbers to build a crib on the ledge, filling the crib with sandbags to provide the bridge sections with a solid foundation.

The Brockway truck drove up and unloaded the treadway bridge sections. Partridge now had everything in place; he had the equipment and he had the people with the necessary experience to install the bridge. First, the treadway bridge sections were laid in place and next, the four inch thick plywood center panels were placed on the lips between the steel treadways.

The bridge repair was finally completed shortly before 4:00 pm. It was just a couple weeks until the shortest day of the year and darkness was only an hour away. Partridge drove to the top of the hill to tell Litzenberg that the repair was completed. Litzenberg said, "Let's get going. Move 'em out!"

It took a while for the heavy equipment at the head of the convoy to slowly wind its way down the steep, snow-covered road to the bridge site.

Unfortunately more bad news soon followed. When Partridge returned to the bridge, he got word that a mile to the south, a steel railroad bridge had been blown and was blocking the road. Moving this steel bridge off the road would be another monumental task.

He looked at one of his engineers and asked, "What in the hell else can go wrong?"

It wouldn't take long to answer the question.

At five o'clock it was very dark and flashlights were needed to guide the vehicles across the bridge. The first elements of the division to arrive were heavy items of equipment including a tractor pulling an eight cubic yard pan used in the construction of the airstrip. When it started across, there was a loud crack and everyone at first thought it was a rifle shot, but it wasn't. The tractor, with one track on the steel treadway and one track on the plywood, broke through the plywood.

Partridge felt like someone had punched him in the stomach. The operator carefully took himself off the tractor and walked back to the road. The bridge was now totally blocked and useless and the 1,400 vehicles of the division were not going anywhere. This could not have happened at a worse time or in a worse place.

Fortunately there was an expert operator nearby. His name was Master Sergeant Wilfred Prosser. He was the foreman in the construction of the airfield at Hagaru. Partridge had come to depend on this man. He got the job done. Partridge called him over and said, "Sergeant, can we get that tractor off the bridge without killing someone?"

"Colonel," said Prosser, "let me have a look."

It was a long, long way to the bottom and it would take a lot of courage to move that tractor off the bridge. Prosser slowly stepped on to the bridge and inspected the broken plywood. He had a flashlight in his hand and he could clearly see the cracked plywood.

When he gingerly returned to the road he told Partridge, "Sir, I think I can move the tractor off the bridge."

"Give it a try, Sergeant, and for God's sake be careful."

Prosser climbed on the tractor; it was still running. His adrenaline was rushing as he skillfully and slowly backed the tractor off the broken plywood and then off the bridge.

Partridge said, "Sergeant, you just saved the day!"

Prosser smiled and said, "Sir, I don't ever want to do something like that again. I was scared shitless."

"You sure could have fooled me," said Partridge. "That took a lot of guts. Great work, Sergeant!"

Partridge was wondering that now with the tractor off the bridge and the plywood useless, how were the division vehicles going to be able to cross the bridge. Some of his men gathered around him and they were all wondering the same thing.

One of his men started to say something and Partridge held up his hand to silence him. The engineering colonel thought he might have a solution and he wanted quiet so he could figure it out. He pulled a pencil and notebook from his jacket. One of his men held a flashlight while he did some calculations. Finally he said, "The narrowest vehicle crossing that bridge will be a jeep. Right?"

"Yes, sir."

"And the widest vehicle a tank?"

"Yes, sir."

Partridge looked at his men and said, "All right, follow me on this: a treadway is forty-five-and-a-half inches wide and the inside tread of a jeep's wheels is forty-five inches. If you add the width of the two treadways that is ninety-one inches, plus forty-five inches is 136 inches, which is two inches wider than the treads of a tank. If we move the treadways forty-five

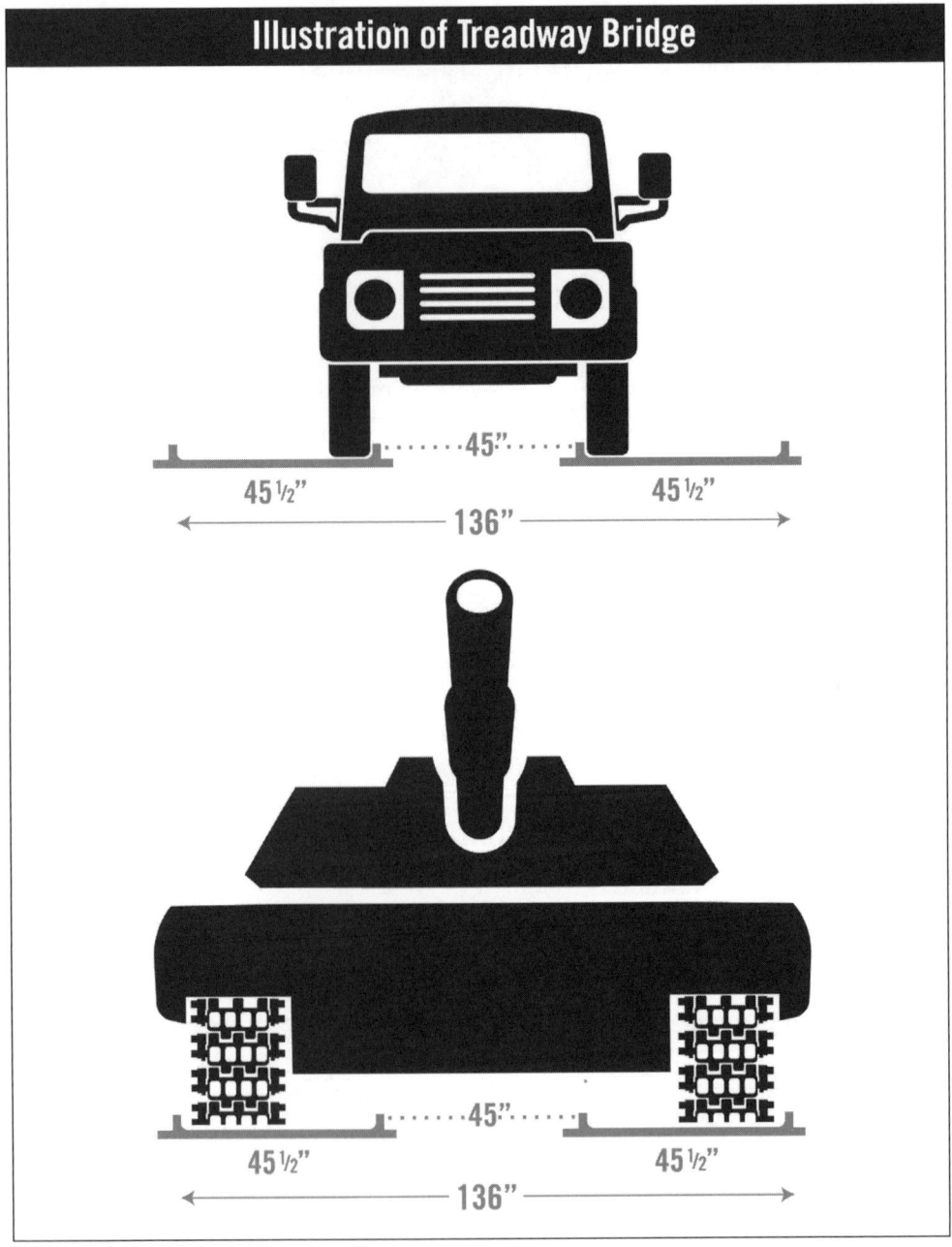

Illustration of Treadway Bridge

45"

45 ½"   45 ½"

136"

45"

45 ½"   45 ½"

136"

inches apart, a jeep can ride the inside lips of the steel treadways. So now the narrowest vehicle, a jeep, can cross, as can the widest, a tank. Let's move these sections."

The Savior

Using a dozer blade an operator pushed the three ton treadways apart and the useless pieces of plywood fell into the black abyss. Next the dozer blade with the help of raw manpower raised one end of the treadway and moved it as close to forty-five inches from the other treadway as possible. A jeep was brought forward to make sure the treadways were exactly forty-five inches apart. After a couple of tries they got it.

Now finally the convoy traffic could resume. The engineers used flash-lights to ensure the safe guidance of all the division's vehicles throughout the night. Progress was slow but steady. One by one the vehicles passed safely over the bridge and back on the road.

Next, the downed steel railroad bridge had to be dealt with. It was only one-half mile from the newly constructed make-shift bridge and it was completely blocking the road. When the engineers and their heavy equipment arrived at the trestle site, there was a lot of discussion as how to remove this massive, steel obstacle. One of the engineers said it would take one hell of a lot of time to blow this apart piece by piece. And time was something they didn't have. No one wanted to be stuck on this road at night with the Chinese army closing in on them.

A lieutenant noticed that a stream had overflowed the road and that the bridge was sitting on this icy surface. He told a bulldozer operator to push against one end of the bridge. When the operator did, the steel skeleton skated across the ice and it swung open like a farm gate, completely clearing the road. Cheers rose from the jubilant engineers. Finally, something went their way. The convoy moved on.

HILL 1081

Earlier in the morning of December 9, Captain Barrow had ordered all his men to test fire their weapons. It was a smart move because many of the rifles and machine guns had frozen in the bitter cold of the night. Barrow's A Company was now ready for the final assault on the crest of Hill 1081.

With the clearing skies, air support was called in to bomb the top of the hill. The bombing, combined with the aggressive assault of Barrow's three rifle platoons forced the Chinese off the crest. By noon, the entire Hill 1081 was in control of the Marines.

But Barrow had paid a heavy price for this real estate: half his men were killed, wounded, or frostbitten. Of the frostbitten Marines, seven would eventually become amputees. Later, records would show that the battle for Hill 1081 would occur on the two coldest days and nights of the entire campaign with temperatures dropping to thirty degrees below zero. The Chinese, however, fared far worse. Barrow's men counted 530 dead Chinese. They found Chinese soldiers frozen to death in their foxholes, as well as those killed by the Marines. The cold was even harder on the poorly equipped Chinese.

KOTO-RI

The day would prove to be another busy and challenging one for General Smith. He not only had a war to deal with, but also a most persistent woman. For the third time, Maggie Higgins appeared on the scene. She had flown in on a casualty evacuation plane and had returned to Koto-ri in spite of Colonel Puller's order to leave. She entered Smith's command post shortly after noon. Smith was busy following the progress of the bridge repair and the troop movement to the south. It was a tense time with so much at stake. He was furious when he saw her come through the door, but he was able to keep his temper under control.

"Miss Higgins," Smith began his reprimand. "I thought you were ordered out of here two days ago by Colonel Puller."

"I was, General Smith," replied Higgins, "but I want to appeal to you personally to let me stay here and march out with the 5th Marines. I made the landing with them at Inchon and I want to walk out with them. General, I am no different from these other male correspondents and I want to be treated exactly as they are. If they can stay, I can stay."

Smith took his pipe from his mouth and said with steely calmness, "Unfortunately, Miss Higgins, that's not going to happen. If you walk down that mountain, you're probably going to get frostbite and some of my Marines are going to have to take care of you. Or what if you get wounded? Or worse yet, what if you get captured? Miss Higgins, I can't spare any of my Marines to babysit you. I need them to fight. You will fly out of here by the end of the day and I want you to know that you will be taking up space that could have been used to evacuate a wounded Marine. That's a consequence of you showing up here."

Smith then turned to his aide and said, "Captain, have Colonel Puller come over here right away."

"Excuse me Miss Higgins, but I have work to do," said Smith dismissively. "Please wait here until Colonel Puller arrives."

Minutes later Puller arrived walking briskly through the flimsy wooden door of the command tent and was surprised to see Maggie Higgins.

Puller barked in his bulldog voice, "What the hell are you doing here? I thought I ordered you to leave."

"You did Colonel, but I thought I would try again."

Puller was furious that she had the nerve to go over his head.

When Smith saw Puller, he walked over and said, "Lewie, I want Miss Higgins out of here by nightfall. Make sure she's on the last plane."

Puller escorted the fuming female correspondent to his CP and assigned another lieutenant to keep an eye on her. "Don't let her out of your sight for even a moment," Puller ordered.

An hour later, Lieutenant General Shepherd, Commanding General of the Fleet Marine Force Pacific, arrived at Smith's CP. He had flown in on a C-47 aircraft. After greetings were exchanged, both men sat down to talk.

"O.P.," said Shepherd, "I saw from the air that the convoy is still a mile away from the bridge. Those vehicles are backed up all the way to Koto-ri."

"General Shepherd, that bridge repair should be completed by early afternoon. Also, I wanted to let you know," continued Smith, "Hill 1081, the high ground south of the bridge has been taken by the Marines. We hope to have all casualties at Koto-ri evacuated by air by the end of the day."

In a statement that surprised Smith, Shepherd said, "O.P., I intend to march out with the division all the way to Chinhung-ni."

It took a lot of backbone for Smith, a two star general, to tell a three star general, his superior, that he wasn't going to walk out. Smith stated emphatically, "No, absolutely not, General Shepherd. I won't allow it. I can't have something happen to you. As you saw when you flew up here this is a very dangerous piece of road. The Chinese aren't finished yet."

Shepherd was not happy with the decision but Smith was the division commander and in charge and that was that.

After meeting with Smith for an hour, Shepherd returned to the airstrip. As the plane was being loaded with the last of the wounded, Puller drove up with Maggie Higgins. They walked over to where Shepherd was standing.

"General," said Puller, "General Smith would appreciate it if Miss Higgins flies out on your plane."

"Maggie," Shepherd said to the reporter, "I'm happy to have you on board."

As darkness fell, the plane took off. Settling into their seats, Shepherd turned to the scowling Higgins and said, "Too bad Maggie, I wanted to walk out with them too, but I understand General Smith's reasons for refusing me. I hope you do as well."

Shortly after Shepherd's plane took off, Bowser informed Smith that the treadway sections had worked. The convoy was now moving across the bridge.

Smith looked at Bowser with an obvious sense of great relief. Thank God! Partridge just delivered another miracle.

A few minutes later, Dr. Hering entered the command tent. Smith was talking to some of the members of his division staff. When he saw the doctor, he waved him over.

Dr. Hering said, "Good evening, General. I'm here to report that all the casualties have been evacuated by air."

"How many, Doctor?" asked the visibly relieved general.

"Sir, we flew out 251 today," said Hering.

"What happens to additional casualties?"

"Well sir, I understand from Colonel Bowser that they will have to be trucked to Hungnam because the airstrip is now closed. With the shrinking perimeter, we can't protect it, although we can still evacuate the most serious casualties by helicopter. Sir, from this tiny airstrip at Koto-ri, we've flown out over 600 casualties."

Smith looked at Hering and said, "Thank you and your doctors and corpsmen for all you have done. You've saved a lot of lives. Hopefully in a couple of days the whole division will be in Hungnam."

"I sure hope so," said Hering. "We are packing up now and getting ready to leave."

The column inched its way forward through the unforgivingly cold night but at a painfully slow pace because with 1,400 vehicles, only one could cross the bridge at a time and the crossing was an agonizing process.

There would continue to be more casualties, both from wounds and from frostbite.

Smith left the command tent shortly before midnight and walked to his quarters. He could feel the cold air on his face and he thought to himself that it seemed colder than ever. As he entered his quarters, he was full of confidence: all the casualties had been evacuated, the 5th and 7th Marines were moving across the bridge and Hill 1081 was in the control of the Marines. He paused to write a short note to his wife. He never liked to write her when the pressure of the situation might bleed through into his letters, but tonight he felt his note would be upbeat. Smith slept well that night.

22

THE TANKS

December 10

KOTO-RI

Early in the morning, there was a lot of activity in the command tent. The movement south during the night had gone well. Clerks were busy packing for the headquarters' move to Hungnam.

Smith grabbed a cup of coffee savoring its warmth on his cold hands. A few minutes later Bowser joined him in the command tent.

"Good morning, General," greeted Bowser. "Good news, sir. The lead elements of the 7th Marines started arriving at Chinhung-ni at 2:45 this morning. Litzenberg thinks he'll have his entire regiment in Chinhung-ni by late afternoon. As soon as they get there, they will be trucked south to the railhead at Majon-dong. There they will board trains for the trip to the port at Hungnam."

"Al, what about casualties last night?"

"Fairly light, General," replied Bowser. "We had ten wounded. We will fly them out this morning by chopper."

Smith took a sip of the steaming coffee and said, "Al, if things continue to go smoothly, let's move the command group to Hungnam mid-morning. The perimeter here is shrinking fast and Puller doesn't need the extra worry about protecting our command post."

Smith returned to his headquarters and started packing the few clothes that he had. Shortly after 10:00 am, Smith and key members of his staff were driven to the landing strip. Smith looked around at the hive of activity involved in moving a command post. It was hard to believe that he had been here for only four days. It seemed like four weeks or four months. Time here was as frozen as the icy ground. Many thousands of men had passed through here. How many were wounded, or frostbitten? He looked around for the last time and then boarded the chopper, which would take him to the rear command post at Hungnam.

When Smith landed he wasted no time. He still had a lot of work to do. He had to evacuate the entire 1st Marine Division from North Korea by sea.

Throughout the day, the 5th Marines followed the 7th Marines into Chinhung-ni. This left only Colonel Puller's 1st Marines holding Koto-ri.

Preparations were made to abandon Koto-ri. Supplies and equipment that couldn't be put on trucks were destroyed and extra ammunition was exploded. The airstrip was closed. There would be more casualties but they would have to walk or ride.

Puller's 1st Marines started pulling out of Koto-ri at 3:00 pm.

When dark came, it appeared that the movement of the 1st Marine Division south would be completed with only minor losses of men and equipment. However, after midnight, two more attacks occurred in areas that the Marines assumed to be safe. The Chinese were not ready to give up.

The first of these took place shortly before 1:00 am as the trucks of the 1st Marines were passing through the village of Sudong. The Chinese stormed out from behind buildings with burp guns and grenades shooting at the truck drivers and setting their vehicles on fire. The Marines responded with a vicious counter-attack pushing back the Chinese. The 1st Marines lost a total of nine trucks and a personnel carrier, with eight men killed and twenty-one wounded. By dawn, the disabled trucks had been cleared and the convoy moved on.

Meanwhile, as planned, the forty Marine tanks were the last vehicles to leave Koto-ri. They would be guarded by a Marine reconnaissance company.

Progress was painfully slow as the big tanks worked their way around the curving, icy road. A mile before the treadway bridge, there was trouble. The brakes on the ninth tank from the rear froze up blocking the road. Unaware of this, the tanks forward continued moving ahead thus stranding the last nine tanks on the road.

With the nine tanks stranded, the Chinese saw their chance and attacked. The battle turned in to a wild fight where it was hard to distinguish friend from foe.

Forty-five minutes later, the tank with the frozen brakes was freed and the tank, followed by the tank behind it, took off once again. The remaining seven tanks were lost to the enemy.

When the last two tanks crossed the bridge, the Marine engineers, as ordered, blew it. The treadway bridge which was the key to the entire breakout and the bridge which had cost so much time and energy, disappeared in a matter of seconds into the darkness below.

23

EMBARKATION

December 11–15

A t 1:00 pm on the eleventh, the last units of the division passed through Chinhung-ni and by 5:30 pm they had cleared Majon-dong. By 9:00 pm, most of the 1st Marine Division units had reached the Hamhung-Hungnam area. A large tent camp greeted the battle-weary Marines on their arrival. The weather on the coast, although still cold, seemed almost balmy compared to the incomprehensible sub-zero temperatures in the mountains.

Upon arriving, they were served hot meals. After nearly fifteen days of continuous fighting with very little food and even less sleep, the Marines of the 1st Division were exhausted physically, mentally and emotionally. The word was passed that all Marines would be evacuated from Hungnam by ship. The foot sore men cheered on hearing this news. Salvation at last!

The embarkation of the 1st Marine Division had already begun with the loading out of Litzenberg's 7th Marines. Murray's 5th Marines would follow on the twelfth with Colonel Puller's 1st Marines boarding the Navy ships on the thirteenth. The Marines' destination would be the port of Pusan in South Korea and from there they would be trucked forty miles to Masan, the new division headquarters.

Late in the morning of the fourteenth, Lieutenant Colonel Partridge stopped in to see General Smith.

"Good morning, sir," greeted the engineer. "Just wanted to let you know that all of your engineering equipment has been loaded on the ships."

Smith was amazed at what these engineers had accomplished. He complimented Partridge again, telling him, "The Roman Empire would never have existed without their engineers, and you and your men deserve the same honors when the history of this war is written."

"Thank you, General," replied Colonel Partridge. "I'll pass your compliments along to the men. I've got a lot of outstanding Marines who made this happen."

Before General Smith boarded his ship which would take him to Pusan, he paid a last visit to the 1st Marine Division cemetery at Hungnam. White crosses covered the small field, each bearing the name of a Marine who had been buried there. The American and Marine Corps flags were lowered to half-mast and the chaplains of three faiths, Catholic, Protestant, and Jewish, said prayers. General Smith, dressed in his winter boots and parka, gave a final tribute. When he was finished a rifle salute rang out and then a lone Marine came forward and put a bugle to his lips and sounded taps.

The last mournful note of taps sounded. Final goodbyes were exchanged between those Marines still standing and their fallen brothers. In a short time, the two month old cemetery went silent; all had left save one: the Commanding General of the 1st Marine Division, O.P. Smith.

A slight breeze started up, unfurling the flags which had been placed at the entrance to the cemetery.

General Smith, now alone with his cap in his hand, looked out over the hundreds of white crosses. A myriad of emotions flooded his thoughts: sorrow, pride, anger, affection all converged and raced through his mind. Each cross belonged to a Marine, one of his Marines and he felt the overwhelming responsibility that only a commanding officer could feel. Flashbacks of similar times in World War II came into his mind. Too many young men, too many young Marines.

His throat tightened and his eyes watered as he started his silent address to his fallen men.

"May God bless and hold each of you," started Smith. "You have given the ultimate sacrifice so that others may live. You are the Marine Corps, you are the eternal Marines. You have suffered and endured hardships that the world cannot comprehend. Rest now men, in God's peace. I salute you."

General Smith put his cap back on and wiped a tear with his glove. He turned and walked back to his jeep where his driver waited.

When Smith sailed on December 15, the curtain went down on one of the most savage battles in the history of the Marine Corps. Out of roughly 22,000 Marines of the 1st Marine Division, 604 were killed, 114 more later died of wounds, 3,485 were wounded, 192 were missing in action and 7,338 were non-battle casualties (mostly frostbite). Close to one half of these battle weary, yet proud, Marines had become casualties.

It was estimated that the Chinese had 25,000 men killed in action and 12,500 wounded from their starting strength of 100,000 soldiers.

Neither numbers nor statistics can ever fully explain or describe the horrors, the brutality or the unrelenting Arctic cold experienced by the Marines of the Chosin Reservoir battle. Nor could it explain the discipline required, valor shown and espirit demonstrated by General Smith's Marines in the cruelest of conditions.

After arriving at Pusan, Smith left the ship and was jeeped to his new headquarters in Masan. Once there he took a pen in hand. In an uncharacteristic show of emotion, General O.P. Smith wrote the following tribute to his men of the 1st Marine Division:

The performance of officers and men in this operation was magnificent. Rarely have all hands in a division participated so intimately in the combat phases of an operation. Every Marine can be justly proud of his participation. In Korea, Tokyo and

Washington there is full appreciation of the remarkable feat of the division. With the knowledge of the determination, professional competence, heroism, devotion to duty, and self-sacrifice displayed by officers and men of this division, my feeling is one of humble pride. No division commander has ever been privileged to command a finer body of men.

EPILOGUE

When Smith arrived in Masan, he told his staff that his goal was to rebuild his division: his men needed rest, hot food, and clean clothing. Equipment had to be repaired and the division needed resupplying. The medical teams screened everyone for frostbite and treated the men for ailments including flu, bronchitis, and intestinal disorders.

The "home by Christmas" predicted by General MacArthur never happened. The Korean War was not yet over; it would drag on for another two-and-a-half years.

Training began again for the Marines and on January 8, they were back in the fight. Smith would command the 1st Marine Division until he was rotated home in April of 1951. On his way home, Smith stopped in Hawaii where he was awarded the Navy Distinguished Service Medal for his outstanding leadership at Chosin Reservoir.

A few weeks later, Major General Smith took over as commanding general at Camp Pendleton, California. His main responsibility was to train Marines for the ongoing Korean War.

On July 1, 1953, Smith took command of the Fleet Marine Force, Atlantic in Norfolk, Virginia, and the following month he was promoted to lieutenant general.

Smith retired in September 1955, and because of his combat record in World War II and the Korean War as well as his exemplary thirty-eight years of service to the Marine Corps, four stars were pinned on his shoulders.

With retirement, a stream of letters flowed in both from senior and junior officers expressing their thanks and respect for his service. He was humbled by their words. But perhaps the greatest praise came from an unexpected source. On September 1, 1955, the *New York Times* wrote the following editorial:

General Oliver P. Smith, United States Marine Corps, retires today at Norfolk, VA, from his last command—Fleet Marine Force, Atlantic—after thirty-eight years of service. Ordinarily the retirement of a general officer attracts little attention. But General Smith is no ordinary general. Scholarly and soft spoken, he looks like a college professor. But he has never lacked the flash of fire; among a Corps noted for its fighting men he was eminent.

He saw bloody fields in World War II—Cape Gloucester; Peleliu; Okinawa—but his name did not become famous until the Korean fighting. The First Marine Division was led ashore at Inchon and toward Seoul, the Korean capital, by General Smith in 1950, and he commanded the division when it was sent by the Tenth Corps deep into the wintry wilderness near the Chosin Reservoir in North Korea.

The sudden Chinese Communist attack with overwhelming numbers and the safe withdrawal of the division to Hungnam and the sea with its dead, its wounded and its equipment—a thirteen-day, seventy mile military epic—epitomized the leadership which General Smith possesses to so high a degree. His remark to a correspondent who mentioned the division's 'retreat'—"Retreat—Hell! We're only attacking in another direction."—has taken its place with the rallying cries of all time.

But the world expects its military leaders to possess physical courage. Much rarer is the quality of moral courage, possessed by Oliver Smith to an unusual degree. It is generally agreed that his leadership saved the First Division at Chosin; it is not generally known that one reason that the division could be saved was that General Smith disobeyed orders. The orders were to continue to advance; the general knew the division was sticking its head into a noose; he ignored the order and consolidated his positions.

Oliver Smith, a Marine's general, richly deserves the honors to be done him this day. The country, proud of his part in her history, will wish him well in retirement.

BIBLIOGRAPHY

BOOKS

Appleman, Lt. Col. Roy E. *Escaping the Trap.* College Station, TX: Texas
    A&M University Press, 1990.

Alexander, Col Joseph H. *The Battle History of the U.S. Marines: A*
    Fellowship of Valor. New York: Harper Perennial, 1997.

Beech, Keyes. *Tokyo and Points East.* Tokyo, Japan: Charles E. Tuttle
    Company, 1955.

Brady, James. *The Marines of Autumn.* New York: St. Martin's Press, 2000.

Camp, Dick. *Leatherneck Legends: Conversations with the Marine Corps'*
    Old Breed. Osceola, WI: Zenith Press, 2006.

Cerasini, Marc. *Heroes: U.S. Marine Corps Medal of Honor Winners.* New
    York: Berkley Books, 2002.

Collier, Peter. *Medal of Honor: Portraits of Valor Beyond the Call of Duty.*
    New York: Artisan, 2003.

Davis, Burke. *Marine! The Life of Chesty Puller.* Boston: Little, Brown, 1962.

Davis, Gen. Raymond G. *The Story of Ray Davis.* Fuquay Varna, NC:
    Research Triangle Publishing, 1995.

Drury, Bob, and Tom Clavin, *The Last Stand of Fox Company.* New York:
    Atlantic Monthly Press, 2009.

Fehrenbach, T.R. *This Kind of War: The Classic Korean War History.*
    Washington, D.C.: Brassey's, 2000.

Hammel, Eric M. *Chosin: Heroic Ordeal of the Korean War.* Novato, CA: Presidio Press, 1981.

Hillenbrand, Laura. *Unbroken.* New York: Random House, 2010.

Higgins, Marguerite. *War in Korea: The Report of a Woman Combat Correspondent.* Garden City, NY: Doubleday & Co. 1951.

Hoffman, Lt. Col Jon T. *Chesty: The Story of Lieutenant General Lewis B. Puller.* New York: Random House, 2001.

Hopkins, William B. *One Bugle, No Drums: The Marine at Chosin Reservoir.* New York: Avon, 1986.

La Bree, Clifton. *The Gentle Warrior: General Oliver Prince Smith, USMC.* Kent, OH: Kent State University Press, 2001.

Lawliss, Chuck. *The Marine Book: A Portrait of America's Military Elite.* New York: Thames and Hudson, 1988.

Leckie, Robert. *The March to Glory.* Cleveland, OH: The World Publishing Co. 1960.

Manchester, William. *American Caesar, Douglas MacArthur 1880–1964.* Boston, MA Little, Brown and Company, 1978.

McClullough, David. *Truman.* New York: Simon & Schuster, 1992.

Montross, Lynn and Canzona, Cap. Nicholas A. USMC. *The Pusan Perimeter Volume I..* Washington D.C.: Historical Branch, G-3, Headquarters at U.S. Marine Corps, 1954.

Montross, Lynn and Canzona, Cap. Nicholas A. USMC. *The Inchon-Seoul Volume II.* Washington D.C.: Historical Branch, G-3, Headquarters at U.S. Marine Corps, 1955.

Montross, Lynn and Canzona, Cap. Nicholas A. USMC. *The Chosin Reservoir Campaign Volume III.* Washington D.C.: Historical Branch, G-3, Headquarters at U.S. Marine Corps, 1957.

Moskin, Robert J. *The U.S. Marine Corps Story.* Boston: Little, Brown, 1992.

O'Donnell, Patrick K. *Give Me Tomorrow: The Korean War's Greatest Untold Story—The Epic Stand of the Marines of George Company.* Cambridge, MA: Da Capo Press, 2010.

Owen, Joseph R. *Colder Than Hell: A Marine Rifle Company at Chosin Reservoir*. Annapolis: Naval Institute, 1996.

Roe, Patrick C. *The Dragon Strikes*. Novato, CA: Presidio Press, 2000.

Russ, Martin. *Breakout: The Chosin Reservoir Campaign, Korea 1950*. New York: Fromm International, 1999.

Shisler, Gail B. *For Country and Corps: The Life of General Oliver P. Smith*. Annapolis, MD: Naval Institute Press, 2009.

Whitcomb, Edgar D. *Escape from Corregidor*. Bloomington, IN: Author House, 1958, 2012.

PAMPHLETS

Alexander, Col Joseph H. *Battle of the Barricades: U.S. Marines in the Recapture of Seoul*. Washington, D.C.: MCHC, 2000.

Simmons, Brig.Gen. Edwin H. *Frozen Chosin: U.S. Marines at the Changjin Reservoir*. Washington, D.C.: MCHC, 2000.

Simmons, Brig.Gen. Edwin H. *Over the Seawall: U.S. Marines at Inchon*. Washington, D.C.: MCHC, 2002.

PERIODICALS

OPS, *Looking Back at Chosin*, Marine Corps Gazette, November 2000.

Bowser, Lt. Gen. Alpha, *Vignette of an Exceptional Marine*, Marine Corps Gazette, December 1995.

Ginchereau, Cap. Eugene H., MC, USNR (Ret.), *RADM Eugene R. Hering Jr.: The Indispensable Military Surgeon*, Navy Medicine Sep-Oct 2003.

PERSONAL PAPERS

Smith, Oliver P., Aide-Memoire box 34, General O.P. Smith collection

OPS notebook, 30 November 1950, box 47, General O.P. Smith collection

ORAL HISTORIES

Beall, Col. Olin L., USMC. Oral History interview by Benis M. Frank, 1970.

Bowser, Lt. Gen. Alpha L., USMC, Oral History interview by Benis M. Frank, 1970.

Smith, Gen. Oliver P., USMC, Oral History interview by Benis M. Frank, 1973.

INTERVIEWS

Craig, Edward A., Assistant Division Commander, 1st Marine Division, Interview by Maj. L.F. Tatem, 1968, Marine Corps Historical Division, Headquarters Marine Corps, Washington, DC.

Partridge, Lt. Colonel J.H., interviewed by Historical Division, HQMC, 25 June 1951.

Smith, Gen. O.P., Interview by Clayton D. James, "Oral reminiscences of General O.P. Smith," box 19, Gen. Oliver P. Smith Collection, Marine Corps Archives and Special Collections, Quantico, VA.

ABOUT THE AUTHORS

NICK RAGLAND graduated from Georgetown University in 1966. He served in Vietnam in 1967–1968 with the Marines as a platoon leader with C Company, 3rd Shore Party Battalion and later as the supply officer of the 1st Battalion 12th Marines. He is chairman of The Gorilla Glue Company. He and his wife, Marty, live in Cincinnati, Ohio. They have five sons and fifteen grandchildren. This is his second book. His first book *Puller's Runner* was published in 2009.

TOM SCHWETTMAN graduated from University of Cincinnati in 1975 and did his graduate work at Miami University (Ohio) in Psychology. He served in the Marine Corps from 1961–1965. He spent most of his career as a sales executive in the medical equipment field and lives in Batesville, Indiana, with his wife Patti. They have five children and nine grandchildren.